AF774615

Achieving ISO/IEC 20000
The differences between BS 15000 and ISO/IEC 20000

The 'Achieving ISO/IEC 20000' series

This publication is the tenth in a series of ten publications related to ISO/IEC 20000. Each publication provides advice on different aspects of ISO/IEC 20000. The books in the 'Achieving ISO/IEC 20000' series are:

Management decisions and documentation (BIP 0030)

Why people matter (BIP 0031)

Making metrics work (BIP 0032)

Managing end-to-end service (BIP 0033)

Finance for service managers (BIP 0034)

Enabling change (BIP 0035)

Keeping the service going (BIP 0036)

Capacity management (BIP 0037)

Integrated service management (BIP 0038)

The differences between BS 15000 and ISO/IEC 20000 (BIP 0039)

This series provides practical guidance and advice on introducing IT service management best practice in accordance with ISO/IEC 20000. More details on the content of each publication are given in *Books in the 'Achieving ISO/IEC 20000' series*, at the end of this book.

Although security issues are covered in ISO/IEC 20000, the 'Achieving ISO/IEC 20000' series does not cover security requirements. Information on security can be found in the BSI publications that are listed in the *Bibliography* in Appendix A.

Other publications

BSI also publishes:

A managers' guide to service management (BIP 0005) is intended for managers who are new to support services or who are faced with major changes to their existing support facility. This book takes the form of informative explanations, guidance and recommendations.

IT service management – Self-assessment workbook (BIP 0015) is an easy to use checklist that complements ISO/IEC 20000 and is designed to assist an organization's internal assessment of their services and the extent to which they conform to the specified requirements in ISO/IEC 20000.

Achieving ISO/IEC 20000
The differences between BS 15000 and ISO/IEC 20000

Dr Jenny Dugmore

Business Information

First published in the UK in 2005

by
BSI
389 Chiswick High Road
London W4 4AL

© British Standards Institution 2005

All rights reserved. Except as permitted under the *Copyright, Designs and Patents Act 1988*, no part of this publication may be reproduced, stored in a retrieval system or transmitted in any form or by any means – electronic, photocopying, recording or otherwise – without prior permission in writing from the publisher.

Whilst every care has been taken in developing and compiling this publication, BSI accepts no liability for any loss or damage caused, arising directly or indirectly in connection with reliance on its contents except to the extent that such liability may not be excluded in law.

The rights of Jenny Dugmore to be identified as the author of this Work have been asserted by her in accordance with sections 77 and 78 of the *Copyright, Designs and Patents Act 1988*.

Typeset in Frutiger by Typobatics Limited
Printed in Great Britain by MPG Books, Bodmin, Cornwall

British Library Cataloguing in Publication Data
A catalogue record for this book is available from the British Library

ISBN 0-580-47348-1

Contents

Foreword

This publication is intended for use by the many organizations and individuals who have used BS 15000 as the basis for audits, self-assessment, training, service improvements and procurement or for requirements of commercial service providers in contracts. It provides a detailed comparison of BS 15000 and ISO/IEC 20000. It includes the reasons for changes and the implications of the changes on the meaning of the standard.

As BS 15000 was withdrawn as a valid standard when ISO/IEC 20000 was published in December 2005, this publication has been produced to provide a detailed mapping of the differences between the two standards, for both Parts 1 and 2. It is intended to help all users of BS 15000 get up to speed on ISO/IEC 20000 as soon as possible, allowing them to make the transition from BS 15000 to ISO/IEC 20000 simply, with low risk and fast.

Dr Jenny Dugmore

Acknowledgements

The work during 2004 and 2005 on re-drafting BS 15000 to become ISO/IEC 20000 has involved many national standards bodies as well as the BSI committee that originally produced BS 15000. I would like to thank them for sharing their views and providing constructive criticism and suggestions.

Particular thanks must go to the BSI committee on IT service management, both past and present, and ISO/IEC JTC 1/SC 7 committee members who have been involved more recently in the re-drafting and plans for the future development of the ISO/IEC 20000 series. These include Jim Clinch, Louise Howson, Alan Miller, Lynda Cooper, David Keech, Don Page, Shirley Lacy, John Groom, Aidan Lawes, Bridget Veitch, Hilary Faul and Ivor Macfarlane, Alison Holt (chair of the SC 7 Ballot resolution meeting), Ogawa Kiyoshi, Darcie Destito and Melanie Cheung.

Finally, I would like to thank Simone Levy and Kieran Parkinson of BSI for their support during the production of the 'Achieving ISO/IEC 20000' series.

Introduction

What is ISO/IEC 20000?

ISO/IEC 20000 is the first IT service management process standard to be produced by the International Organization for Standardization (ISO), and is based on the knowledge and experience gained by experts working in the field. ISO/IEC 20000 was produced by Technical Committee ISO/IEC JTC 1/SC 7, *Software and system engineering*, and was based on BS 15000, which was produced by BSI Technical Committee BDD/3, *Information services management*.

ISO/IEC 20000 is in two parts.

- ISO/IEC 20000-1 is a specification containing requirements that must be met in order to achieve ISO/IEC 20000.
- ISO/IEC 20000-2 is a code of practice on how to achieve the requirements in ISO/IEC 20000-1.

The requirements in Part 1 (i.e. ISO/IEC 20000-1) are applicable to service providers of all sizes and types, regardless of whether the organization is public or private sector, internal or external. The recommendations in Part 2 (i.e. ISO/IEC 20000-2) are optional approaches to achieving the requirements in Part 1. Although optional, the recommendations are also practical and proven methods that are normally appropriate.

The purpose of ISO/IEC 20000

ISO/IEC 20000 provides the basis for assessing whether service providers have best practice, reliable, repeatable and measurable processes applied consistently across their organization. As a process-based standard the requirements are independent of organizational structure or of the tools used to automate the service management processes. ISO/IEC 20000-1 provides the basis for formal certification schemes and other audits.

The 'Achieving ISO/IEC 20000' series

The first nine books of the 'Achieving ISO/IEC 20000' series are designed to explain the requirements of ISO/IEC 20000. Table 1 provides a clause-by-clause guide to the content of each of the books in the 'Achieving ISO/IEC 20000' series.

This tenth publication in the 'Achieving ISO/IEC 20000' series explains the differences between both parts of the two standards, assisting with the transition from BS 15000 to ISO/IEC 20000.

Additional advice

Service providers aiming for ISO/IEC 20000 may find it useful to seek advice on best practice, the qualifications that are available for individual service management professionals and ISO/IEC 20000 certification. Details of these can be found via the web pages in Appendix A.

Table 1 – Clause-by-clause guide to the 'Achieving ISO/IEC 20000' series

ISO/IEC 20000 clause	BIP 0030	BIP 0031	BIP 0032	BIP 0033	BIP 0034	BIP 0035	BIP 0036	BIP 0037	BIP 0038	BIP 0039
Terms and definitions	■									■
Management responsibility	■								■	■
Documentation requirements	■									■
Competence, awareness and training		■								■
Planning and implementing service management									■	■
Plan – Do – Check – Act cycle									■	■
Planning and implementing new or changed services									■	■
Service level management				■						■
Service reporting			■							■
Service continuity and availability management							■			■
Budgeting and accounting for IT services					■					■
Information security management	See BSI publications on information security management in the BIP 0070 series									■
Capacity management								■		■
Business relationship management				■						■
Supplier management				■						■
Incident management							■			■
Problem management							■			■
Configuration management						■				■
Change management						■				■
Release management						■				■

CHAPTER 1

What difference does it make?

Why does ISO/IEC 20000 matter?

ISO/IEC 20000 matters because high standards of service management are important to service providers and to those reliant on their services. ISO/IEC 20000 is the international standard for service management. It applies to an industry that can only be truly effective if it operates to the same high standard internationally. ISO/IEC 20000 is the distillation of the service management industry's best practices, giving service providers a set of tangible and objective goals to aim for and achieve. The service management industry, which is typically dependent on a complex supply chain, needs a common standard for best practice service management processes, combined with general best practice management processes.

ISO/IEC 20000 provides a vehicle for the following.

- **Industry recognition** that adopting the standards in ISO/IEC 20000 delivers better services, continuing improvements, greater cost-effectiveness and higher customer satisfaction.
- **International recognition** that the service management best practices specified in ISO/IEC 20000 are directly linked to the majority of the IT spend, with figures such as '80% of IT spend' quoted. This represents a significant proportion of the gross national/domestic product of technologically advanced nations. It reflects recognition that achieving ISO/IEC 20000 will bring benefits at national level, above and beyond being able to specify compliance as part of the procurement process.
- **Recognition by senior IT managers** that achieving ISO/IEC 20000 means they must 'walk the talk', as their leadership makes or breaks the ability of the whole service provider to achieve best practice.
- **Understanding** that ISO/IEC 20000 isn't only about a series of service management processes; it's about management best practices in service provision, including integration of service management with other management processes and with continual improvements.
- **Adoption of an international standard for IT service management by more organizations,** in particular those who have been waiting for an international version of BS 15000. This is particularly so for any organization that spans national boundaries and those who must have a high standard of service management in order to meet governance and regulatory requirements. This applies to both large and small organizations.

Why has this book been produced?

There are many organizations and individuals that have adopted the use of BS 15000, which was superseded by ISO/IEC 20000 when it was published; ISO/IEC 20000 is a direct replacement for BS 15000.

The users of BS 15000 did not just span those service providers that have achieved BS 15000 via a formal certification scheme. There are many organizations that are using BS 15000 as a goal for service improvement, even if they do not intend to be formally audited under an external and independent scheme. There are also organizations who require their service providers to have achieved BS 15000, and include this requirement in tender documents or contracts. These organizations need to understand the differences between BS 15000 and ISO/IEC 20000 when changing requirements in tender documents and contracts to reflect the publication of ISO/IEC 20000. Other organizations include audit, consulting and training companies. There are also many individuals that have been trained in the understanding and use of BS 15000.

This book is intended for use by:

- service provider's:
  - senior management (including the senior responsible owner of service management);
  - process owners;
  - internal auditors;
  - service improvement teams;
  - practitioners;
- organizations that have waited for an international version of BS 15000;
- the audit industry's:
  - external auditors;
  - owners/managers of certification schemes;
  - audit industry regulators;
- national standards bodies, particularly those who have published a national version of BS 15000;
- consulting and training companies;
- developers of best practice service management advice and guidance;
- tool vendors who wish to use ISO/IEC 20000 as a source of advice for the design of their tools intended for use in automation and support of service management processes.

This book has been produced to provide a smooth transition to using ISO/IEC 20000.

What is being compared in this book?

This publication compares BS 15000:2002 to BS ISO/IEC 20000:2005 (the UK national implementation of ISO/IEC 20000). BS ISO/IEC 20000 has **identical** requirements and recommendations to ISO/IEC 20000, but has additional information in the National Foreword and National Annex. This information includes references to the continuing role of the BSI committee on IT service management in support for the new international standard, references to how more information may be obtained and details of publications including the 'Achieving ISO/IEC 20000' series and the IT Infrastructure library (ITIL®[1]).

As the Part 1 requirements and Part 2 recommendations are identical in both BS ISO/IEC 20000 and ISO/IEC 20000, the National Foreword and National Annex do not alter the impact or status of the standard, and therefore have no the impact on the service provider with regards to any certification schemes. Permission to use any of the copyright material in ISO/IEC 20000 may also be obtained from BSI, as was the case for BS 15000.

A brief history of ISO/IEC 20000

The BS 15000 series has evolved over several years, and ultimately became ISO/IEC 20000 after a fast-tracking conversion process.

The BS 15000 standard on service management became more detailed and more exacting as the understanding of service management best practices developed. Starting as a minority interest it has expanded to a widely accepted discipline. The second edition of BS 15000 was expanded to include management system requirements following consultation with early adopters of the first edition. The alignment between BS 15000 and ITIL has continued with ISO/IEC 20000. The milestones in the publication of ISO/IEC 20000 are shown in Table 2.

[1] ITIL (IT infrastructure library) is a registered trade mark of OGC (the Office of Government Commerce), Rosebery Court, St. Andrew's Business Park, Norwich, Norfolk, NR7 0HS. This information is given for the convenience of users of this standard and does not constitute an endorsement by BSI of this product.

Table 2 – Milestones in the history of BS 15000 and ISO/IEC 20000

Year	Milestone
1989	Committee responsible for service management established by BSI.
1995	First code of practice (4 service management processes) published by BSI.
1998	Second code of practice standard (13 service management processes) published by BSI.
2000	Self-assessment workbook (new editions produced in later years) published by BSI.
	First edition of the Managers' guide to service management published by BSI (new editions produced in later years).
	First edition of the specification and revised code of practice published by BSI.
	Early Adopters scheme set up and managed by the BSI committee on IT service management.
2001	Recommendations from Early Adopters reviewed and adopted during the production of the second edition of BS 15000 published by BSI.
2002	Second edition of the specification, including new management system requirements published by the BSI.
2003	June – Certification scheme launched by itSMF.
2004	February – First three companies certified against BS 15000.
	Planning started for fast-track submission to ISO/IEC JTC 1/SC 7.
	November – submission made to ISO/IEC JTC 1/SC 7.
2005	May – 'Yes' vote on fast track adoption from national standards bodies.
	May-October – Comment resolution and planning of new work by ISO/IEC JTC 1/SC 7 project editor.
	November – Editorial work by ISO editor.
	New work item proposal for the development of ISO/IEC 20000 submitted by BSI to ISO/IEC JTC 1/SC 7.
	December – Publication of ISO/IEC 20000 and withdrawal of BS 15000.

The fast track process

Fast-tracking a national standard to become an ISO or ISO/IEC standard is done when it is believed that a standard produced by a single national standards body, like the BSI, has relevance to many nations. BS 15000 was a national standard being used internationally. This was the reason for the decision to submit BS 15000 to JTC 1 in November 2004 as a fast track candidate to become ISO/IEC 20000.

The voting announced in May 2005 was in favour of adoption of BS 15000 as ISO/IEC 20000. Votes were accompanied by comments that were discussed and resolved, ending in a final meeting in October 2005 that agreed changes to the draft ISO/IEC 20000. Generally, fast tracking a standard to become an ISO/IEC standard means that the changes required to the document are not fundamental.

Another outcome of the comment resolution was a set of suggestions for future work on ISO/IEC 20000. The new work currently being planned is part of the normal standards development process managed by ISO/IEC JTC 1/SC 7. This is separate from the fast-track process that finished with the publication of ISO/IEC 2000 in December 2005.

It was also agreed that the standard would be managed by the newly established Work Group 25 (WG25), which is part of ISO/IEC JTC 1/SC 7 *Software and system engineering*, as described in the section on *The future of ISO/IEC 20000.*

The future of ISO/IEC 20000

Proposals

Timetable and priorities of the new work item will be influenced by the report of an ISO/IEC JTC 1/SC 7 Study Group on IT Operations: Service Management and Delivery, ITC Governance. The publication date for this is Spring 2006.

Key topics that are under consideration for the long term future of ISO/IEC 20000 at the time of publication of this book can be summarized as:

- clarification/agreement of applicability and scope;
- improvements and extension of the recommendations in Part 2;
- commitment to SC 7's work on harmonization and vocabulary, while retaining alignment with widely adopted service management best practice material; and
- longer term development of requirements in Part 1.

Applicability and scope of ISO/IEC 20000

Suggestions on applicability include extension to explicitly include communications or for the scope to become the much wider 'technology enabled services'.

Scoping the processes for service management planning and service improvements and in particular scope statements for certification audits have all been discussed. It has been proposed that the scope of ISO/IEC 20000 is reviewed by Work Group 25, with the aim of reaching agreement on the most appropriate definition for the applicability of the standard and scope (including scope statements for certification audits). Options include additional clauses in the existing Parts 1 and 2, but also the development of a Part 3, covering scoping/applicability/audit guidelines.

Part 2 – Detail and style

Comments and suggestions have been made on the desirability of the next edition of ISO/IEC 20000-2 (the code of practice) being extended and improved. This work could be done in parallel with and complement the work on applicability and scope described above. This is expected to take the form of:

- **consistency** in the level of detail in Part 2, relative to the detail on Part 1;
- **more recommendations** that reflect the interests of an international audience;
- **style of advice** to be targeted more for use by junior management and practitioners rather than senior managers
- **planning and process integration** to cover additional advice on topics such as planning for service management and service improvements, the integration of processes and in particular the integration of service management processes with the other management processes (i.e. the integration of processes in clauses 3 to 5 with processes in clauses 6 to 10).

Harmonization/vocabulary

Discussions on the future of ISO/IEC 20000 have been on the basis of there being commitment to harmonization of international standards, whilst also retaining the beneficial aspects of alignment between ISO/IEC 20000 and the widely used ITIL. The proposals for the enhancements to ISO/IEC 20000 are therefore based on the expectation that harmonization will be actively supported by those who will be involved in Work Group 25. Specific topics that have been identified include terminology, process reference models and alignment between ISO/IEC 20000 and ITIL.

Terminology

Currently, most terms in ISO/IEC 20000 are used with the commonly accepted English language dictionary definition. Special terms are mainly definitions based on ITIL. A small number are based on those in other international standards (i.e. 'documents' and 'records'). It is proposed that Work Group 25 considers the implications of ITIL terms and the work of the SC 7 Vocabulary group.

Process reference models

The inclusion of objectives in ISO/IEC 20000 does not align to other SC 7 process-based standards.. This will be considered by Work Group 25, including the suggestion that ISO/IEC 20000 should conform to ISO/IEC 15504-2, *Requirements for process reference models*.

Alignment between ISO/IEC 20000 and ITIL

The alignment between BS 15000 and ITIL has continued as alignment between ISO/IEC 20000 and ITIL. At the date of publication of ISO/IEC 20000, ITIL is being reviewed and revised. It is expected that the benefits of keeping the two aligned will be a feature of both ITIL revisions and of any future changes to ISO/IEC 20000.

CHAPTER 2

BS 15000 and ISO/IEC 20000 Part 1

Introduction

Relatively few changes have been made to the requirements (Part 1) during re-drafting of BS 15000 to become ISO/IEC 20000. Differences were primarily clarification of requirements and standardization of terms, such as 'service provider', 'supplier' etc. Table 3 shows the differences in structure and objectives of Part 1. All differences are relative to BS 15000, e.g. subclause 2.15 was not included in BS 15000, and so that cell is blank. It is included in ISO/IEC 20000, as is shown. Words that have been changed are underlined. Details of the changes are given in Chapter 6.

Table 3 – BS 15000 and ISO/IEC 20000 Part 1

BS 15000-1	ISO/IEC 20000-1
	National Foreword
Foreword	Foreword
Introduction	**Introduction**
1 Scope	**1 Scope**
2 Terms and definitions	**2 Terms and definitions**
availability	2.1 availability
2.2 baseline	2.2 baseline
2.3 change record	2.3 change record
2.4 configuration item (CI)	2.4 configuration item (CI)
2.5 configuration management database (CMDB)	2.5 configuration management database (CMDB)
2.6 document	2.6 document
2.7 incident	2.7 incident
2.8 problem	2.8 problem
2.9 record	2.9 record
2.10 release	2.10 release
2.11 request for change	2.11 request for change
2.12 service desk	2.12 service desk
2.13 service level agreement (SLA)	2.13 service level agreement (SLA)
2.14 service management	2.14 service management
	2.15 service provider

BS 15000-1	ISO/IEC 20000-1
3 Requirements for a mangement system	**3 Requirements for a mangement system**
Objective: To provide a management system, including policies and a framework to enable the effective management and implementation of all IT services.	*Objective: To provide a management system, including policies and a framework to enable the effective management and implementation of all IT services.*
3.1 Management responsibility	3.1 Management responsibility
3.2 Documentation requirements	3.2 Documentation requirements
3.3 Competence, awareness and training	3.3 Competence, awareness and training
4 Planning and implementing service management	**4 Planning and implementing service management**
4.1 Plan service management (Plan)	4.1 Plan service management (Plan)
Objective: To plan the implementation and delivery of service management.	*Objective: To plan the implementation and delivery of service management.*
4.2 Implement service management and provide the services (Do)	4.2 Implement service management and provide the services (Do)
Objective: To implement the service management objectives and plan.	*Objective: To implement the service management objectives and plan.*
4.3 Monitoring, measuring and reviewing (Check)	4.3 Monitoring, measuring and reviewing (Check)
Objective: To monitor, measure and review that the service management objectives and plan are being achieved.	*Objective: To monitor, measure and review that the service management objectives and plan are being achieved.*
4.4 <u>Continuous</u> improvement (Act)	4.4 <u>Continual</u> improvement (Act)
Objective: To improve the effectiveness and efficiency of service delivery and management	*Objective: To improve the effectiveness and efficiency of service delivery and management*
	4.4.1 Policy
	4.4.2 Management of improvements
	4.4.3 Activities
5 Planning and implementing new or changed services	**5 Planning and implementing new or changed services**
Objective: To ensure that new services and changes to services will be deliverable and manageable at the <u>right</u> cost and service quality.	*Objective: To ensure that new services and changes to services will be deliverable and manageable at the <u>agreed</u> cost and service quality.*
6 Service delivery process	**6 Service delivery process**
6.1 Service level management	6.1 Service level management
Objective: To define, agree, record and manage levels of service.	*Objective: To define, agree, record and manage levels of service.*
6.2 Service reporting	6.2 Service reporting
Objective: To produce agreed, timely, reliable, accurate reports for informed decision making and effective communication.	*Objective: To produce agreed, timely, reliable, accurate reports for informed decision making and effective communication.*

BS 15000-1	ISO/IEC 20000-1
6.3 Availability and service continuity management	6.3 Service continuity and availability management
Objective: To ensure that agreed obligations to customers can be met in all circumstances.	*Objective: To ensure that agreed service continuity and availability commitments to customers can be met in all circumstances*
6.4 Budgeting and accounting for IT services	6.4 Budgeting and accounting for IT services
Objective: To budget and account for the cost of service provision.	*Objective: To budget and account for the cost of service provision.*
6.5 Capacity management	6.5 Capacity management
Objective: To ensure that the organization has, at all times, sufficient capacity to meet the current and future agreed demands of the business.	*Objective: To ensure that the service provider has, at all times, sufficient capacity to meet the current and future agreed demands of the customer's business needs.*
6.6 Information security management	6.6 Information security management
Objective: To manage information security effectively within all service activities.	*Objective: To manage information security effectively within all service activities.*
7 Relationship processes	**7 Relationship processes**
	7.1 General
7.1 Business relationship management	7.2 Business relationship management
Objective: To establish and maintain a good relationship between the service provider and the customer based on understanding the customer and their business drivers.	*Objective: To establish and maintain a good relationship between the service provider and the customer based on understanding the customer and their business drivers.*
7.2 Supplier management	7.3 Supplier management
Objective: To manage third party suppliers to ensure the provision of seamless, quality services.	*Objective: To manage suppliers to ensure the provision of seamless, quality services.*
8 Resolution processes	**8 Resolution processes**
	8.1 Background
8.1 Incident management	8.2 Incident management
Objective: To restore agreed service to the business as soon as possible or to respond to service requests.	*Objective: To restore agreed service to the business as soon as possible or to respond to service requests.*
8.2 Problem management	8.3 Problem management
Objective: To minimize disruption to the business by proactive identification and analysis of the cause of service incidents and by managing problems to closure.	*Objective: To minimize disruption to the business by proactive identification and analysis of the cause of incidents and by managing problems to closure.*
9 Control processes	**9 Control processes**
9.1 Configuration management	9.1 Configuration management
Objective: To define and control the components of the service and infrastructure and maintain accurate configuration information.	*Objective: To define and control the components of the service and infrastructure and maintain accurate configuration information.*

BS 15000-1	ISO/IEC 20000-1
9.2 Change management	9.2 Change management
Objective: To ensure all changes are assessed, approved, implemented and reviewed in a controlled manner.	*Objective: To ensure all changes are assessed, approved, implemented and reviewed in a controlled manner.*
10 Release process	**10 Release process**
10.1 Release management	10.1 Release management process
Objective: To deliver, distribute and track one or more changes in a release into the live environment.	*Objective: To deliver, distribute and track one or more changes in a release into the live environment.*
Bibliography	**Bibliography**
Standards publications	
Non-standards publications	
	National annex NA (informative)
	National bibliography

CHAPTER 3

BS 15000 and ISO/IEC 20000 Part 2

Introduction

This chapter maps the differences in the structure and objectives of Part 2. All differences are relative to BS 15000, e.g. the heading 'Why is part 2 needed?' is shown in the left hand column (3rd row down) but was deleted during re-drafting and so the equivalent in the headings in ISO/IEC 20000 is blank.

The sequence of processes in clause 5 of BS 15000-2 placed service reporting after service continuity and availability management. In ISO/IEC 20000 this sequence has been reversed. The sequence now matches that in BS 15000-1 and ISO/IEC 20000-1. Table 4 shows subclauses 6.2 and 6.3 out of numeric order, as the comparison is to BS 15000-2. Other differences are shown by underlining, with more detail provided in Chapter 7.

Many of the differences are the addition of new subclause headings. This groups the recommendations within each clause and subclause, which meets the ISO/IEC formatting requirements, and allows unique references based on the subclause heading and the paragraph or letter number of the alpha lists which are used extensively.

Table 4 – BS 15000 and ISO/IEC 20000 Part 2

BS 15000-2	ISO/IEC 20000-2
	National Foreword
Introduction	**Introduction**
Why is part 2 needed?	
1 Scope	**1 Scope**
2 Terms and definitions	**2 Terms and definitions**
3 The management system	**3 The management system**
Objective: To provide a management system, including policies and a framework to enable the effective management and implementation of all IT services.	*Objective: To provide a management system, including policies and a framework to enable the effective management and implementation of all IT services.*
3.1 Management responsibility	3.1 Management responsibility
3.2 Documentation requirements	3.2 Documentation requirements
3.3 Managing documents	
3.4 Competence, awareness and training	3.3 Competence, awareness and training
<u>*Objective: To ensure that service management personnel are competent to undertaken their role.*</u>	

BS 15000-2	ISO/IEC 20000-2
3.4.1 Introduction	3.3.1 General
	3.3.2 Professional development
	3.3.3 Approaches to be considered
4 Planning and implementing service management	**4 Planning and implementing service management**
4.1 Plan service management (Plan)	4.1 Plan service management (Plan)
Objective: To plan the implementation and delivery of service management.	*Objective: To plan the implementation and delivery of service management.*
	4.1.1 Scope of service management
	4.1.2 Planning approaches
	4.1.3 Events to be considered
	4.1.4 Scope and contents of the plan
4.2 Implement service management and provide the services (Do)	4.2 Implement service management and provide the services (Do)
Objective: To implement the service management plan and achieve the defined service objectives.	*Objective: To implement the service management objectives and plan.*
4.3 Monitoring, measuring and reviewing (Check)	4.3 Monitoring, measuring and reviewing (Check)
Objective: To monitor, measure and review that the service management objectives and plan are being achieved.	*Objective: To monitor, measure and review that the service management objectives and plan are being achieved.*
4.3.1 Measurement and analysis	
4.4 Continuous improvement (Act)	4.4 Continual improvement (Act)
Objective: To improve the effectiveness and efficiency of service delivery and management.	*Objective: To improve the effectiveness and efficiency of service delivery and management.*
4.4.1 Policy	4.4.1 Policy
4.4.2 Planning for service improvements	4.4.2 Planning for service improvements
4.4.3 Planning and implementing new or changed services	**5 Planning and implementing new or changed services**
Objective: To ensure that new services and changes to services will be deliverable and manageable at the agreed cost and service quality.	*Objective: To ensure that new services and changes to services will be deliverable and manageable at the agreed cost and service quality.*
	5.1 Topics for consideration
	5.2 Change records
5 Service delivery processes	**6 Service delivery processes**
5.1 Service level management	6.1 Service level management
Objective: To define, agree, record, and manage levels of service.	*Objective: To define, agree, record, and manage levels of service.*
5.1.1 Service catalogue	6.1.1 Service catalogue
5.1.2 Service level agreements (SLAs)	6.1.2 Service level agreements (SLAs)

BS 15000-2	ISO/IEC 20000-2
5.1.3 Service level management (SLM) process	6.1.3 Service level management (SLM) process
5.1.4 Supporting service agreements	6.1.4 Supporting service agreements
<u>5.2</u> Service continuity and availability management	<u>6.3</u> Service continuity and availability management
Objective: To ensure that agreed obligations to customers can be met <u>in all circumstances from normal through to a major loss of service</u>.	*Objective: To ensure that agreed service continuity and availability commitments to customers can be met <u>in all circumstances</u>.*
<u>5.2.1</u> General	<u>6.3.1</u> General
<u>5.2.2</u> Availability monitoring and activities	<u>6.3.2</u> Availability monitoring and activities
<u>5.2.3</u> Service continuity strategy	<u>6.3.3</u> Service continuity strategy
<u>5.2.4</u> Service continuity planning and testing	<u>6.3.4</u> Service continuity planning and testing
<u>5.3</u> Service reporting	<u>6.2</u> Service reporting
Objective: To produce agreed timely, reliable, accurate reports for informed decision making and effective communication.	*Objective: To produce agreed timely, reliable, accurate reports for informed decision making and effective communication.*
<u>5.3.1</u> Policy	<u>6.2.1</u> Policy
<u>5.3.2</u> Purpose and quality checks on service reports	<u>6.2.2</u> Purpose and quality checks on service reports
<u>5.3.3</u> Service reports	<u>6.2.3</u> Service reports
5.4 Budgeting and accounting for IT services	6.4 Budgeting and accounting for IT services
Objective: To budget and account for the cost of service provision.	*Objective: To budget and account for the cost of service provision.*
	6.4.1 General
5.4.1 Policy	6.4.2 Policy
5.4.2 Budgeting	6.4.3 Budgeting
5.4.3 Accounting	6.4.4 Accounting
5.5 Capacity management	6.5 Capacity management
Objective: To ensure that the <u>organization</u> has, at all times, sufficient capacity to meet the current and future agreed <u>demands of the business</u>.	*Objective: To ensure that the <u>service provider</u> has, at all times, sufficient capacity to meet the current and future agreed <u>demands of customer's business needs</u>.*
5.6 Information security management	6.6 Information security management
Objective: To manage information security effectively within all service activities.	*Objective: To manage information security effectively within all service activities.*
	6.6.1 General
5.6.1 Identifying and classifying information assets	6.6.2 Identifying and classifying information assets
5.6.2 Security risk <u>assessment</u>	6.6.3 Security risk <u>assessment practices</u>

BS 15000-2	ISO/IEC 20000-2
	6.6.4 Risks to information assets
	6.6.5 Security and availability of information
5.6.3 Controls	6.6.6 Controls
	6.6.7 Documents and records
6 Relationship processes	**7 Relationship processes**
6.1 General	7.1 General
6.2 Business relationship management	7.2 Business relationship management
Objective: To establish and maintain a good relationship between the service provider and customer, based on understanding the customer and their business drivers.	*Objective: To establish and maintain a good relationship between the service provider and customer based on understanding the customer and their business drivers.*
6.2.1 Service reviews	7.2.1 Service reviews
6.2.2 Service complaints	7.2.2 Service complaints
6.2.3 Customer satisfaction measurement	7.2.3 Customer satisfaction measurement
6.3 Supplier management	7.3 Supplier management
Objective: To manage the service provider(s) to ensure the provision of seamless, quality services.	*Objective: To manage supplier(s) to ensure the provision of seamless, quality services.*
6.3.1 Introduction	7.3.1 Introduction
6.3.2 Contract management	7.3.2 Contract management
6.3.3 Service definition	7.3.3 Service definition
6.3.4 Managing multiple suppliers	7.3.4 Managing multiple suppliers
6.3.5 Contractual disputes management	7.3.5 Contractual disputes management
6.3.6 Contract end	7.3.6 Contract end
7 Resolution processes	**8 Resolution processes**
7.1 Background	8.1 Background
7.1.1 Setting priorities	8.1.1 Setting priorities
7.1.2 Workarounds	8.1.1 Workarounds
7.2 Incident management	8.2 Incident management
Objective: To restore normal service as soon as possible in order to minimize business disruption.	*Objective: To restore agreed service to the business as soon as possible or to respond to service requests.*
	8.2.1 General
7.2.1 Major incidents	8.2.2 Major incidents

BS 15000-2	ISO/IEC 20000-2
7.3 Problem management	8.3 Problem management
Objective: To identify and manage the underlying causes of service incidents whilst minimizing or preventing disruption to the customers.	*Objective: To minimize disruption to the business by proactive identification and analysis of the causes of incidents and by managing problems to closure.*
7.3.1 Scope of problem management	8.3.1 Scope of problem management
7.3.2 Initiation of problem management	8.3.2 Initiation of problem management
7.3.3 Known errors	8.3.3 Known errors
7.3.4 Problem resolution	8.3.4 Problem resolution
7.3.5 Communication	8.3.5 Communication
7.3.6 Tracking and escalation	8.3.6 Tracking and escalation
7.3.7 Incident and problem record closure	8.3.7 Incident and problem record closure
7.3.8 Problem reviews	8.3.8 Problem reviews
	8.3.9 Topics for reviews
7.3.9 Problem prevention	8.3.10 Problem prevention
8 Control processes	**9 Control processes**
8.1 Configuration management	9.1 Configuration management
Objective: To define and control the components of the service and infrastructure and maintain accurate configuration information.	*Objective: To define and control the components of the service and infrastructure and maintain accurate configuration information.*
8.1.1 Configuration management planning and implementation	9.1.1 Configuration management planning and implementation
8.1.2 Configuration identification	8.1.2 Configuration identification
8.1.3 Configuration control	9.1.3 Configuration control
8.1.4 Configuration status accounting and reporting	9.1.4 Configuration status accounting and reporting
8.1.5 Configuration verification and audit	9.1.5 Configuration verification and audit
8.2 Change management	9.2 Change management
Objective: To ensure all changes are assessed, approved, implemented and reviewed in a controlled manner.	*Objective: To ensure all changes are assessed, approved, implemented and reviewed in a controlled manner.*
8.2.1 Planning and implementation	9.2.1 Planning and implementation
8.2.2 Closing and reviewing the change request	9.2.2 Closing and reviewing the change request
8.2.3 Emergency changes	9.2.3 Emergency changes
8.2.4 Change management reporting, analysis and actions	9.2.4 Change management reporting, analysis and actions

BS 15000-2	ISO/IEC 20000-2
9 Release management processes	**10 Release process**
	10.1 Release management process
Objective: To deliver, distribute and track one or more changes into the live environment.	*Objective: To deliver, distribute and track one or more changes into the live environment.*
	10.1.1 General
9.2 Release policy	10.1.2 Release policy
9.3 Release and roll-out planning	10.1.3 Release and roll-out planning
9.4 Developing or acquiring software	10.1.4 Developing or acquiring software
9.5 Design, build and configure release	10.1.5 Design, build and configure release
9.6 Release verification and acceptance	10.1.6 Release verification and acceptance
	10.1.7 Documentation
9.7 Roll-out, distribution and installation	10.1.8 Roll-out, distribution and installation
9.8 Post release and roll-out	10.1.9 Post release and roll-out
Bibliography	**Bibliography**
National Annex	**National Annex**
National Bibliography	**National Bibliography**

CHAPTER 4

Comparison of Parts 1 and 2

Introduction

Table 5 maps the differences between Parts 1 and 2 of ISO/IEC 20000. The structure of Parts 1 and 2 are more closely aligned than in BS 15000. For example, the clause on planning or implementing new or changed services is clause 5 in both parts. Similarly, service reporting is covered in clause 6.2 and service continuity and availability management are covered in clause 6.3 in both parts.

Table 5 can be used to compare the contents of both the specification and the code of practice, for headings and objectives.

Some third level headings have been added to Part 2 to allow cross-reference by quoting a heading number and an alpha list. These third level headings have not been introduced into Part 1 as the clauses in the specification are much shorter and do not need as much subdivision.

Table 5 – Comparison of Parts 1 and 2

ISO/IEC 20000-1 Specification	ISO/IEC 20000-2 Code of Practice
National Foreword	National Foreword
Foreword	Foreword
Introduction	Introduction
1 Scope	1 Scope
2 Terms and definitions	2 Terms and definitions
2.1 availability	Cross-reference to Part 1.
2.2 baseline	
2.3 change record	
2.4 configuration item (CI)	
2.5 configuration management database (CMDB)	
2.6 document	
2.7 incident	
2.8 problem	
2.9 record	
2.10 release	
2.11 request for change	

ISO/IEC 20000-1 Specification	ISO/IEC 20000-2 Code of Practice
2.12 service desk	
2.13 service level agreement (SLA)	
2.14 service management	
2.15 service provider	
3 Requirements for a management system	**3 The management system**
Objective: To provide a management system, including policies and a framework to enable the effective management and implementation of all IT services.	*Objective: To provide a management system, including policies and a framework to enable the effective management and implementation of all IT services.*
3.1 Management responsibility	**3.1 Management responsibility**
3.2 Documentation requirements	**3.2 Documentation requirements**
3.3 Competence, awareness and training	**3.3 Competence, awareness and training**
	3.3.1 General
	3.3.2 Professional development
	3.3.3 Approaches to be considered
4 Planning and implementing service management	**4 Planning and implementing service management**
4.1 Plan service management (Plan)	**4.1 Plan service management (Plan)**
Objective: To plan the implementation and delivery of service management.	*Objective: To plan the implementation and delivery of service management.*
	4.1.1 Scope of service management
	4.1.2 Planning approaches
	4.1.3 Events to be considered
	4.1.4 Scope and contents of the plan
4.2 Implement service management and provide the services (Do)	**4.2 Implement service management and provide the services (Do)**
Objective: To implement the service management objectives and plan.	*Objective: To implement the service management objectives and plan.*
4.3 Monitoring, measuring and reviewing (Check)	**4.3 Monitoring, measuring and reviewing (Check)**
Objective: To monitor, measure and review that the service management objectives and plan are being achieved.	*Objective: To monitor, measure and review that the service management objectives and plan are being achieved.*
4.4 Continual improvement (Act)	**4.4 Continual improvement (Act)**
Objective: To improve the effectiveness and efficiency of service delivery and management.	*Objective: To improve the effectiveness and efficiency of service delivery and management.*
4.4.1 Policy	4.4.1 Policy
4.4.2 Management of improvements	4.4.2 Planning for service improvements

ISO/IEC 20000-1 Specification	ISO/IEC 20000-2 Code of Practice
4.4.3 Activities	
5 Planning and implementing new or changed services	**5 Planning and implementing new or changed services**
Objective: To ensure that new services and changes to services will be deliverable and manageable at the agreed cost and service quality.	*Objective: To ensure that new services and changes to services will be deliverable and manageable at the agreed cost and service quality.*
	5.1 Topics for consideration
	5.2 Change records
6 Service delivery process	**6 Service delivery processes**
6.1 Service level management	**6.1 Service level management**
Objective: To define, agree, record and manage levels of service.	*Objective: To define, agree, record, and manage levels of service.*
	6.1.1 Service catalogue
	6.1.2 Service level agreements (SLAs)
	6.1.3 Service level management (SLM) process
	6.1.4 Supporting service agreements
6.2 Service reporting	**6.2 Service reporting**
Objective: To produce agreed, timely, reliable, accurate reports for informed decision making and effective communication.	*Objective: To produce agreed, timely, reliable, accurate reports for informed decision making and effective communication.*
	6.2.1 Policy
	6.2.2 Purpose and quality checks on service reports
	6.2.3 Service reports
6.3 Service continuity and availability management	**6.3 Service continuity and availability management**
Objective: To ensure that agreed service continuity and availability commitments to customers can be met in all circumstances	*Objective: To ensure that agreed service continuity and availability commitments to customers can be met in all circumstances.*
	6.3.1 General
	6.3.2 Availability monitoring and activities
	6.3.3 Service continuity strategy
	6.3.4 Service continuity planning and testing
6.4 Budgeting and accounting for IT services	**6.4 Budgeting and accounting for IT services**
Objective: To budget and account for the cost of service provision.	*Objective: To budget and account for the cost of service provision.*
	6.4.1 General
	6.4.2 Policy
	6.4.3 Budgeting
	6.4.4 Accounting

ISO/IEC 20000-1 Specification	ISO/IEC 20000-2 Code of Practice
6.5 Capacity management	**6.5 Capacity management**
Objective: To ensure that the service provider has, at all times, sufficient capacity to meet the current and future agreed demands of the customer's business needs.	*Objective: To ensure that the service provider has, at all times, sufficient capacity to meet the current and future agreed demands of customer's business needs.*
6.6 Information security management	**6.6 Information security management**
Objective: To manage information security effectively within all service activities.	*Objective: To manage information security effectively within all service activities.*
	6.6.1 General
	6.6.2 Identifying and classifying information assets
	6.6.3 Security risk assessment practices
	6.6.4 Risks to information assets
	6.6.5 Security and availability of information
	6.6.6 Controls
	6.6.7 Documents and records
7 Relationship processes	**7 Relationship processes**
7.1 General	**7.1 General**
7.2 Business relationship management	**7.2 Business relationship management**
Objective: To establish and maintain a good relationship between the service provider and the customer based on understanding the customer and their business drivers.	*Objective: To establish and maintain a good relationship between the service provider and customer, based on understanding the customer and their business drivers.*
	7.2.1 Service reviews
	7.2.2 Service complaints
	7.2.3 Customer satisfaction measurement
7.3 Supplier management	**7.3 Supplier management**
Objective: To manage suppliers to ensure the provision of seamless, quality services.	*Objective: To manage suppliers to ensure the provision of seamless, quality services.*
	7.3.1 Introduction
	7.3.2 Contract management
	7.3.3 Service definition
	7.3.4 Managing multiple suppliers
	7.3.5 Contractual disputes management
	7.3.6 Contract end

ISO/IEC 20000-1 Specification	ISO/IEC 20000-2 Code of Practice
8 Resolution processes	**8 Resolution processes**
8.1 Background	**8.1 Background**
	8.1.1 Setting priorities
	8.1.2 Workarounds
8.2 Incident management	**8.2 Incident management**
Objective: To restore agreed service to the business as soon as possible or to respond to service requests.	*Objective: To restore agreed service to the business as soon as possible or to respond to service requests.*
	8.2.1 General
	8.2.2 Major incident
8.3 Problem management	**8.3 Problem management**
Objective: To minimize disruption to the business by proactive identification and analysis of the cause of incidents and by managing problems to closure.	*Objective: To minimize disruption to the business by proactive identification and analysis of the causes of incidents and by managing problems to closure.*
	8.3.1 Scope of problem management
	8.3.2 Initiation of problem management
	8.3.3 Known errors
	8.3.4 Problem resolution
	8.3.5 Communication
	8.3.6 Tracking and escalation
	8.3.7 Incident and problem record closure
	8.3.8 Problem reviews
	8.3.9 Topics for reviews
	8.3.10 Problem prevention
9 Control processes	**9 Control processes**
9.1 Configuration management	**9.1 Configuration management**
Objective: To define and control the components of the service and infrastructure and maintain accurate configuration information.	*Objective: To define and control the components of the service and infrastructure and maintain accurate configuration information.*
	9.1.1 Configuration management planning and implementation
	8.1.2 Configuration identification
	9.1.3 Configuration control
	9.1.4 Configuration status accounting and reporting
	9.1.5 Configuration verification and audit

ISO/IEC 20000-1 Specification	ISO/IEC 20000-2 Code of Practice
9.2 Change management	**9.2 Change management**
Objective: To ensure all changes are assessed, approved, implemented and reviewed in a controlled manner.	*Objective: To ensure all changes are assessed, approved, implemented and reviewed in a controlled manner.*
	9.2.1 Planning and implementation
	9.2.2 Closing and reviewing the change request
	9.2.3 Emergency changes
	9.2.4 Change management reporting, analysis and actions
10 Release process	**10 Release process**
10.1 Release management process	**10.1 Release management process**
Objective: To deliver, distribute and track one or more changes in a release into the live environment.	*Objective: To deliver, distribute and track one or more changes into the live environment.*
	10.1.1 General
	10.1.2 Release policy
	10.1.3 Release and roll-out planning
	10.1.4 Developing or acquiring software
	10.1.5 Design, build and configure release
	10.1.6 Release verification and acceptance
	10.1.7 Documentation
	10.1.8 Roll-out, distribution and installation
	10.1.9 Post release and roll-out
Bibliography	**Bibliography**
National annex NA (informative)	**National annex NA (informative)**
National Bibliography	**National Bibliography**

Chapter 5

Coding of changes

Introduction

The differences between BS 15000 and ISO/IEC 20000 are covered in detail in Tables 6 and 7 later in this book. This chapter explains how the changes to both parts of BS 15000 are classified in these tables, based on the significance of the change to current users of BS 15000.

Tables 6 and 7 are made up of eight columns. The first two columns contain a comparison of the clauses in BS 15000 and ISO/IEC 20000. The next five columns describe the type of change that has been made, and are classified as follows:

1. N/C – no change;
2. Form – formatting change;
3. Edit – editorial change/correction;
4. Rec – recommendation change;
5. Req – requirement change.

These columns are explained in greater detail and with examples in *Explanation of column headings in Tables 6 and 7 indicating changes between BS 15000 and ISO/IEC 20000*. The changes to each clause are indicated by a tick ✔ in the appropriate column. The last column in Tables 6 and 7 contain additional information.

Changes in text up to but not including clause 2 do not affect requirements as this text provides background and is primarily standard text in every ISO/IEC standard. However, some of this text has been shown where it was considered to be of particular interest to the transition from BS 15000 to ISO/IEC 20000.

Unchanged text is not shown in Tables 6 and 7, unless it is required to give context to requirements or recommendations that have changed. Where a whole clause or subclause has not changed, only the heading is included.

Text that has been changed is normally underlined. Where whole sentences or paragraphs of text are new or have been completely deleted there is only text in one of the columns. This makes the change easy to see and so this is not underlined. Additional headings or deleted headings are not underlined for the same reason.

Explanation of column headings in Tables 6 and 7 indicating changes between BS 15000 and ISO/IEC 20000

N/C

This indicates that there has been no change. Unchanged text is not normally shown in this publication, except where it is required for the context of a change that is shown.

Font changes and the separation of sentences into smaller paragraphs are not classed as a change and are therefore included in this category.

Some minor changes (e.g. punctuation, minor typing mistakes etc) are also in this category. These are changes where it is very unlikely there would be any confusion on the implications of the change and there is no risk of the change being misunderstood and/or treated as a change to requirements or recommendations.

Form

This type of change is a formatting change, and makes no difference to the original spirit and intent of the requirements or recommendations. Most of these changes have arisen from differences in the rules for the structure and drafting of standards that applied to BS 15000 (i.e. BS-0) and ISO/IEC 20000, (i.e. ISO/IEC Directives, Part 2).

There are over 300 of these changes in Parts 1 and 2. Most of these are re-numbering, use of alpha lists and insertion of subclause headings. Many of these changes have been made to make the standard easier to use and to follow, including cross-references.

Many of the changes that align the structure of Parts 1 and 2 are also in this category, such as ISO/IEC 20000-2, 4.4.3 *Planning and implementing new or changed services* being re-numbered as clause 5, which aligns Part 2 to Parts 1.

These changes are mainly of interest to any user of BS 15000 who has relied on the structure and detail of BS 15000 for audit guidelines, training material or for plans for service management and service improvements.

Edit

These changes are editorial and have been made mainly to make the standard easier to understand by re-wording to avoid ambiguity of meaning. This also makes the standard easier to interpret for an international audience, particularly those who do not speak English as their first language.

Other examples of editorial changes include standardizing on the use of terms such as service provider, supplier, customer, organization, user. Some of these requirements and recommendations may have been misinterpreted, so the reader's attention is drawn to these even though the original intent of the clause is unchanged. There are approximately 140 of these changes in Parts 1 and 2.

Editorial changes also include changes made to how plans are referred to. For example, 'the service improvement plan' has been changed to '*a plan for improving the service*'. This change was made partly to standardize how plans are referred to but also to avoid the requirements and recommendations being interpreted as the need for a special plan with the title '*Service Improvement Plan*', which have arisen because of the use of the definite article. The intention of BS 15000 was that the choice of a plan, several plans, a full scale programme and how each plan or programme was to be named be left to the individual service provider, as long as the planning was effective and the requirements are met as a result of the planning. Hence the use of the indefinite article and re-wording.

For this type of change the principle adopted is that if the new wording is still aligned to the original intent of the BSI committee that produced BS 15000, this change type is indicated. If the new wording has changed the requirement or recommendation originally intended, one of the next two changes ('Rec' or 'Req') are indicated.

Rec

These are changes to the recommendations in ISO/IEC 20000. There are approximately 35 of these changes in Parts 1 and 2. The majority of these changes have been made to the code of practice, but some have occurred as a NOTE in the Part 1. This is because a NOTE in Part 1, which is the specification, has the same status as clause in Part 2, which is the code of practice, i.e. it does **not** form part of the requirements and can be ignored if a service provider chooses to do so. The service provider aiming to achieve ISO/IEC 20000 can also ignore the Part 2 recommendations. However, these are tried and tested methods that are normally applicable and will be of interest to most organizations.

Req

This is the most significant category as it represents a small number of changes to requirements, i.e. those things that are compulsory. There are 16 of these changes in Part 1. The changes are made up of extensions to original requirements, the insertion of new requirements or the removal of requirements that have been converted into a NOTE.

Some new requirements, e.g. the requirements for service reporting in Chapter 7, have arisen from small changes to wording. For example, in service reporting the use of the words '*Typically service reports shall include.....*' was ambiguous as a requirement as the word '*Typically....* ' is not appropriate for a specification as it potentially allows a service provider to not produce the reports specified but to still claim compliance under the guise of atypical circumstances. Changing the requirement to '*Service reporting shall....*' has clarified the intention of this requirement'. Another example has arisen with the change from '*Complaint procedure...*' to '*Complaints process...* ' in ISO/IEC 20000-1, 7.2. This is a change to requirements and an ISO/IEC 20000 auditor now has to check only the complaints process, and may choose not to check the procedure recommended by ISO/IEC 20000-2.

CHAPTER 6

Detailed changes to Part 1

Introduction

This chapter describes the changes to the requirements in detail. The column headings described in detail in Chapter 5 are used to distinguish between the different types of changes to Part 1.

To make it easier to identify a NOTE in Table 6, NOTEs are included as serif fonts, with the rest of the clause text as san serif font. This is because a NOTE in Part 1 (the specification) is significant **as it does not form part of the requirements**. A NOTE provides advice, recommendations or an explanation, and is similar in status to the whole of Part 2 (the code of practice).

Table 6 – The Part 1 differences

BS 15000-1	ISO/IEC 20000-1	N\C	Form	Edit	Rec.	Req.	Commentary
IT Service Management – Part 1: Specification for service management	Information technology – Service management – Part 1: Specification			✔			Change to the title to reflect normal ISO/IEC naming practices.
	National Foreword		✔				New clause; does not change the requirements.
	This British Standard reproduces verbatim ISO/IEC 20000-1:2005 and implements it as the UK national standard.		✔				New text.
	It supersedes BS 15000-1:2002 which is withdrawn.		✔				New text.
	The UK participation in its preparation was entrusted by Technical Committee IST15/-/8, IT service management, which has the responsibility to: — aid enquirers to understand the text; — present to the responsible international/European committee any enquiries on the interpretation, or proposals for change, and keep UK interests informed; — monitor related international and European developments and promulgate them in the UK.		✔				Moved from the original BS 15000 Foreword and reworded to reflect current BSI committee structure.
	A list of organizations represented on this committee can be obtained on request to its secretary.		✔				New text.
	BS ISO/IEC 20000 is based on the knowledge and experience gained by experts working in the field. It can be used in conjunction with a number of best practice publications, listed in the national annex NA.		✔				Moved from Foreword and reworded to reflect new status and the addition of a National Annex.

Table 6 – The Part 1 differences (continued)

BS 15000-1	ISO/IEC 20000-1	N/C	Form	Edit	Rec.	Req.	Commentary
	Cross references The British Standards which implement international publications referred to in this document may be found in the BSI catalogue under the section entitled "International Standards Correspondence Index", or by using the "Search" facility of the BSI Electronic Catalogue or of British Standards Online.		✔				New text.
	This publication does not purport to include all the necessary provisions of a contract.		✔				Moved from the original BS 15000 Foreword.
	Users are responsible for its correct application.		✔				Moved from the original BS 15000 Foreword.
	Compliance with a British Standard does not of itself confer immunity from legal obligations.		✔				Moved from the original BS 15000 Foreword.
	Foreword		✔				
	ISO (the International Organization for Standardization) and IEC (the International Electrotechnical Commission) form the specialized system for worldwide standardization.		✔				Explanation of the acronym ISO in ISO/IEC 20000. Explanation of the acronym IEC in ISO/IEC 20000.
	National bodies that are members of ISO or IEC participate in the development of International Standards through technical committees established by the respective organization to deal with particular fields of technical activity.		✔				Standard ISO/IEC text.
	ISO and IEC technical committees collaborate in fields of mutual interest.		✔				

Table 6 – The Part 1 differences (continued)

BS 15000-1	ISO/IEC 20000-1	N\C	Form	Edit	Rec.	Req.	Commentary
	Other international organizations, governmental and non-governmental, in liaison with ISO and IEC, also take part in the work.		✔				Standard ISO/IEC text.
	In the field of information technology, ISO and IEC have established a joint technical committee ISO/IEC JTC 1.		✔				
	ISO and IEC have established a joint technical committee ISO/IEC JTC 1.		✔				Explanation of JTC 1.
	International Standards are drafted in accordance with the rules given in the ISO/IEC Directives, Part 2.		✔				
	The main task of the joint technical committee is to prepare International Standards.		✔				
	Draft International Standards adopted by the joint technical committee are circulated to national bodies for voting.		✔				
	Publication as an International Standard requires approval by at least 75% of the national bodies casting a vote.		✔				New ISO/IEC standard text.
	Attention is drawn to the possibility that some of the elements of this document may be the subject of patent rights.		✔				
	ISO and IEC shall not be held responsible for identifying any or all such patent rights.		✔				

Table 6 – The Part 1 differences (continued)

BS 15000-1	ISO/IEC 20000-1	N/C	Form	Edit	Rec.	Req.	Commentary
BS 15000-1 has been prepared by Technical Committee BDD/3 and is based on the knowledge and experience gained by experts working in the field.	ISO/IEC 20000-1 was prepared by BSI (as BS15000 1) and was adopted, under a special "fast-track procedure", by Joint Technical Committee ISO/IEC JTC 1, *Information technology*, in parallel with its approval by national bodies of ISO and IEC.		✔				Reworded to be suitable for an international standard. More information on the BSI's role in this is included in the National Foreword.
It supersedes BS 15000:2000, which is withdrawn.			✔				
BS 15000 is issued in two parts: — Part 1: Specification for service management; — Part 2: Code of practice for service management.	ISO/IEC 20000 consists of the following parts, under the general title Information Technology — Service management: - Part 1: Specification - Part 2: Code of Practice			✔			Change to title during formatting due to ISO/IEC conventions on naming. This does not change the scope, requirements or recommendations.
This British Standard should be used in conjunction with BS 15000-2, the code of practice associated with this specification.			✔				
It may also be used in conjunction with DISC PD 0005, Management Overview, and DISC PD 0015, IT Service Management — Self-assessment Workbook. DISC PD 0015 is a checklist that complements this specification. This workbook has been designed to assist organizations to assess the extent to which their IT services conform to the specified requirements.			✔				Deleted as national references are not appropriate in international standards.

Table 6 – The Part 1 differences (continued)

BS 15000-1	ISO/IEC 20000-1	N\C	Form	Edit	Rec.	Req.	Commentary
A standard forms part of a trade description when cited by number or when compliance with it is claimed.			✔				Deleted. This is not a change to the requirements or recommendations.
Even if a standard does not specifically require marking with its number, organizations often refer to it in documentation.			✔				Deleted. This is not a change to the requirements or recommendations.
Such marking or reference constitutes a unilateral claim that the product, process or service conforms to all the relevant requirements of the standard cited.			✔				Deleted. This is not a change to the requirements or recommendations.
The person making the claim is legally responsible in the UK under the Trade Descriptions Act 1968 [1].			✔				Deleted as reference to national legislation is not appropriate for an international standard.
It is assumed that the execution of the provisions of this standard is entrusted to appropriately qualified and competent people.			✔				Moved to Introduction.
A **British** Standard does not purport to include all necessary provisions of a contract.			✔				Moved to Introduction, minor re-wording to reflect international status.
Also included in the National Foreword. Users of **British** Standards are responsible for their correct application.			✔				
Compliance with a **British** Standard does not of itself confer immunity from legal obligations.			✔				

Table 6 – The Part 1 differences (continued)

BS 15000-1	ISO/IEC 20000-1	N/C	Form	Edit	Rec.	Req.	Commentary
Attention is drawn to the following statutory regulations: — Trade Descriptions Act 1968 [1]; — Data Protection Act 1998 [2]; — Computer Misuse Act 1990 [3]; — Copyright, Designs and Patent Act 1988 [4].			✔				Deleted as references to national legislation are not appropriate for an international standard.
BSI Committee BDD/3, whose constitution is shown in this British Standard, takes collective responsibility for its preparation under the authority of the Standards Board.			✔				Details replaced by details of the SC 7 committee.
The Committee wishes to acknowledge the personal contributions of: The Committee wishes to acknowledge the personal contributions of: Jenny Dugmore, Service Matters Ltd Lynda Cooper, Xansa, Ivor Evans, IT service Improvement Ltd, Hilary Faul, BBC Technology Ltd. John Groom, Office of Government Commerce (OGC), Shirley Lacy, ConnectSphere Ltd and representative of the BCS, Aidan Lawes, IT Service Management Forum (itSMF), Ivor Macfarlane, IT service Improvement Ltd, Don Page, Marval Ltd.			✔				Details deleted as not appropriate under formatting rules.

Table 6 – The Part 1 differences (continued)

BS 15000-1	ISO/IEC 20000-1	N\C	Form	Edit	Rec.	Req.	Commentary
Introduction	**Introduction**	✔					Title unchanged, text modified.
<u>**This standard promotes**</u> the adoption of an integrated process approach to effectively deliver managed services to meet the business and customer requirements.	<u>**This part of ISO/IEC 20000 promotes**</u> the adoption of an integrated process approach to effectively deliver managed services to meet the business and customer requirements.		✔				Change to how the standard is referred to in line with normal practices for ISO/IEC standard.
For an organization to function effectively it has to identify and manage numerous linked activities.	For an organization to function effectively it has to identify and manage numerous linked activities.	✔					No change, left in for context of other changes.
An activity using resources, and managed in order to enable the transformation of inputs into outputs, can be considered as a process.	An activity using resources, and managed in order to enable the transformation of inputs into outputs, can be considered as a process.	✔					
Often the output from one process forms an input to another.	Often the output from one process forms an input to another.	✔					
Co-ordinated integration and implementation of the service management processes provides the ongoing control, greater efficiency and opportunities for <u>**continuous**</u> improvement.	Co-ordinated integration and implementation of the service management processes provides the ongoing control, greater efficiency and opportunities for <u>**continual**</u> improvement.			✔			This is correction of the word '*continuous*' by use of '*continual*' in line with other management system standards. This is not a change to the intentions of this clause, nor is it part of the requirements.
Performing the activities and processes requires people in the service desk, service support, service delivery and operations teams to be well organized and co-ordinated.	Performing the activities and processes requires people in the service desk, service support, service delivery and operations teams to be well organized and co-ordinated.	✔					No change, left in for context of other changes.
Appropriate tools are also required to ensure that the processes are effective and efficient.	Appropriate tools are also required to ensure that the processes are effective and efficient.	✔					No change, left in for context of other changes.

Table 6 – The Part 1 differences (continued)

BS 15000-1	ISO/IEC 20000-1	N/C	Form	Edit	Rec.	Req.	Commentary
	It is assumed that the execution of the provisions of this part of ISO/IEC 20000 is entrusted to appropriately qualified and competent people.		✔				Moved from original BS 15000 Foreword.
	An International Standard does not purport to include all necessary provisions of a contract.		✔				
	Users of International Standards are responsible for their correct application.		✔				Moved from original BS 15000 Foreword, minor re-wording to reflect international status.
	Compliance with an International Standard does not of itself confer immunity from legal obligations.		✔				
NOTE The methodology known as "Plan-Do-Check-Act" (PDCA) can be applied to all processes. PDCA can be described as follows: 1) **Plan**: establish the objectives and processes necessary to deliver results in accordance with customer requirements and the organization's policies; 2) **Do**: implement the processes; 3) **Check**: monitor and measure processes and services against policies, objectives and requirements and report the results; 4) **Act**: take actions to continually improve process performance.			✔				Moved to clause 4, there is no change to wording. It is not a change to the requirements even though this text has been moved to clause 4. This is because a NOTE is advice, recommendations or explanations, not part of the requirements. A NOTE can be ignored by the service provider aiming to achieve ISO/IEC 20000.

Table 6 – The Part 1 differences (continued)

BS 15000-1	ISO/IEC 20000-1	N\C	Form	Edit	Rec.	Req.	Commentary
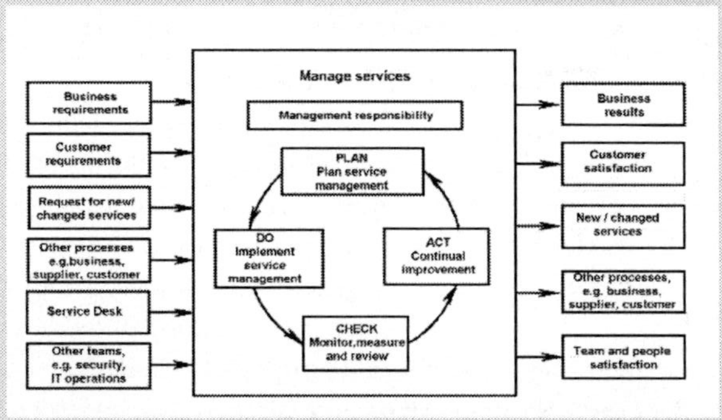			✔				Moved to clause 4, this is not a change to wording or requirements
Figure 1 — Plan-Do-Check-Act methodology for service management processes			✔				Moved to clause 4, this is not a change to wording or requirements
The model shown in Figure 1 illustrates the process and process linkages presented in clauses **5, 6 and 7.**				✔			Moved to clause 4, minor change to the clause cross-referenced, but not to requirements.
1 Scope	**1 Scope**	✔					Title unchanged, text in the scope has been modified.
This specification defines the requirements for an **organization** to deliver managed services of an acceptable quality for its customers.	**This part of ISO/IEC 20000** defines the requirements for a **service provider** to deliver managed services of an acceptable quality for its customers.			✔			Change to how the standard is referred to in line with normal practices for ISO/IEC standard. There is also clarification from standardization on the use of service provider.

Table 6 – The Part 1 differences (continued)

BS 15000-1	ISO/IEC 20000-1	N/C	Form	Edit	Rec.	Req.	Commentary
It may be used:	It may be used:	✔					No change, left in for context of other changes.
a) by businesses that are going out to tender for their services;	a) by businesses that are going out to tender for their services;	✔					
b) by businesses that require a consistent approach by all service providers in a supply chain;	b) by businesses that require a consistent approach by all service providers in a supply chain;	✔					
c) by service providers to benchmark their IT service management;	c) by service providers to benchmark their IT service management;	✔					
d) as the basis for **<u>an assessment;</u>**	d) as the basis for an **<u>independent assessment;</u>**			✔			Reference to independent assessment rather than assessment and deletion of the reference to the certification scheme, as this is not appropriate in an ISO/IEC standard. This is not a change to the requirements or recommendations.
….**<u>which may lead to a formal certification;</u>**				✔			
e) by an organization who needs to demonstrate the ability to provide services that meet customer requirements; and	e) by an organization which needs to demonstrate the ability to provide services that meet customer requirements; and	✔					No change, left in for context of other changes.
f) by an organization which aims to improve service through the effective application of processes to monitor and improve service quality.	f) by an organization which aims to improve service through the effective application of processes to monitor and improve service quality.	✔					No change, left in for context of other changes.

Table 6 – The Part 1 differences (continued)

BS 15000-1	ISO/IEC 20000-1	N\C	Form	Edit	Rec.	Req.	Commentary
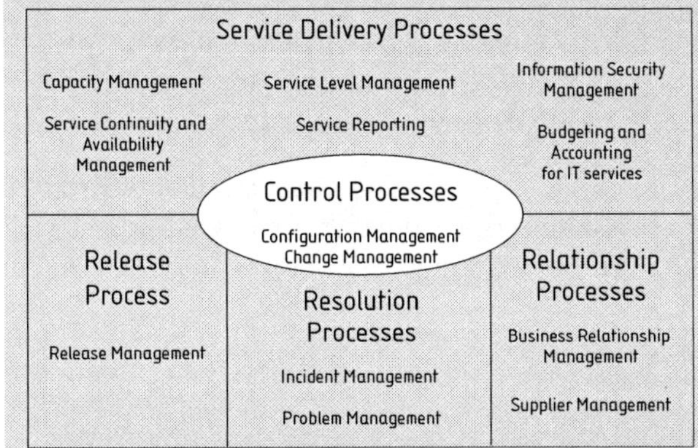	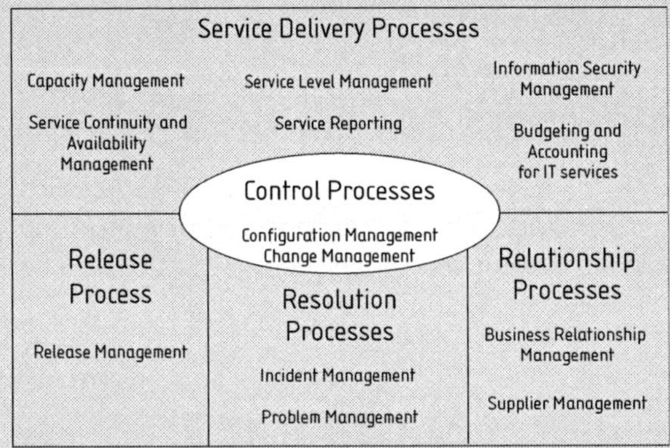	✔					No change, left in for context of other changes.
Figure 2 — Service management processes	**Figure 1— Service management processes**		✔				Change to Figure number due to change in the sequence of the Figures.
This standard specifies a number of closely related service management processes, as shown in Figure 1.	This part of ISO/IEC 20000 specifies a number of closely related service management processes, as shown in Figure 1.		✔				No change, left in for context of other changes.
The relationships between the processes depend on the application within an organization and are generally too complex to model and therefore relationships between processes are not shown in this diagram.	The relationships between the processes depend on the application within an organization and are generally too complex to model and therefore relationships between processes are not shown in this diagram.	✔					No change, left in for context of other changes.
The list of objectives and controls contained in this specification are not exhaustive, and an organization may consider that additional objectives and controls are necessary to meet their particular business needs.	The list of objectives and controls contained in **this part of ISO/IEC 20000** are not exhaustive, and an organization may consider that additional objectives and controls are necessary to meet their particular business needs.		✔				

Table 6 – The Part 1 differences (continued)

BS 15000-1	ISO/IEC 20000-1	N/C	Form	Edit	Rec.	Req.	Commentary
The nature of the business relationship between the service provider and business will determine how the requirements in this Standard are implemented in order to meet the overall objective.	The nature of the business relationship between the service provider and business will determine how the requirements in **this part of ISO/IEC 20000** are implemented in order to meet the overall objective.		✔				No change, left in for context of other changes.
	As a process based standard this part of ISO/IEC 20000 is not intended for product assessment. However, organizations developing service management tools, products and systems may use both this part of ISO/IEC 20000 and **the code of practice** to help them develop tools, products and systems that support best practice service management.			✔			Additional wording on the purpose and scope of the standard. This is not part of the requirements or recommendations.
2 Terms and definitions	**2 Terms and definitions**	✔					Title unchanged, text modified.
For the purposes of **this part of BS 15000**, the following terms and definitions apply.	For the purposes of **this document** the following terms and definitions apply.		✔				Change due to ISO formatting conventions.
availability	**2.1 availability**		✔				Missing clause number included.
ability of a component or service to perform its required function at a stated instant or over a stated period of time	ability of a component or service to perform its required function at a stated instant or over a stated period of time	✔					No change, left in for context of other changes.
NOTE Availability is usually expressed as a ratio of the time that the service is actually available for use by the business to the agreed service hours.	NOTE Availability is usually expressed as a ratio of the time that the service is actually available for use by the business to the agreed service hours.	✔					No change, left in for context of other changes.
2.2 baseline	**2.2 baseline**	✔					
snapshot of the state of a service or **an individual configuration item…**	snapshot of the state of a service or **individual configuration items**…		✔				Clarification / correction of text, not a change to the definition or the requirements.

Table 6 – The Part 1 differences (continued)

BS 15000-1	ISO/IEC 20000-1	N\C	Form	Edit	Rec.	Req.	Commentary
....at a point in time	.... at a point in time (see 2.4)		✔				Cross-reference added as 'configuration item' has not been defined before subclause 2.2.
2.3 change record	**2.3 change record**	✔					No change, left in for context of other changes.
2.4 configuration item (CI)	**2.4 configuration item (CI)**	✔					No change, left in for context of other changes.
2.5 configuration management database (CMDB)	**2.5 configuration management database (CMDB)**	✔					No change, left in for context of other changes.
database containing all the relevant details of each configuration **items** and details of the important relationships between them	database containing all the relevant details of each configuration **item** and details of the important relationships between them		✔				Clarification / correction of text, not change to the definition or the requirements.
2.6 document	**2.6 document**	✔					No change, left in for context of other changes.
information in readable form, including computer data, which is created or received and maintained as evidence of the service provider's intentions with regards to service management	**information and its supporting medium**			✔			Change to definition to align with this term defined in other standards. No change to the intent of the definition or associated requirements.
NOTE 1 In this standard, **records** are distinguished from documents by the fact that they function as evidence of activities, rather than evidence of intentions.	NOTE 1 In this standard, **records (see 2.9)** are distinguished from documents by the fact that they function as evidence of activities, rather than evidence of intentions.		✔				Cross reference added as 'record' has not been defined prior to clause 2.6.
NOTE 2 Examples of documents include policy statements, plans, procedures, service level agreements and contracts.	NOTE 2 Examples of documents include policy statements, plans, procedures, service level agreements and contracts.	✔					No change, left in for context of other changes.

Table 6 – The Part 1 differences (continued)

BS 15000-1	ISO/IEC 20000-1	N/C	Form	Edit	Rec.	Req.	Commentary
2.7 incident	**2.7 incident**	✔					No change, left in for context of other changes.
2.8 problem	**2.8 problem**	✔					No change, left in for context of other changes.
2.9 record	**2.9 record**	✔					No change to heading, left in for context of other changes.
<u>**information in readable form, including computer data, which is created or received and maintained by the service provider as evidence of the performance of service management activities**</u>	<u>**document stating results achieved or providing evidence of activities performed**</u>			✔			Change to definition to align with this term defined in other standards. No change to the intent of the definition or associated requirements.
NOTE 1 In this standard, records are distinguished from documents by the fact that they function as evidence of activities, rather than evidence of intentions.	NOTE 1 In this standard, records are distinguished from documents by the fact that they function as evidence of activities, rather than evidence of intentions.	✔					No change, left in for context of other changes.
NOTE 2 Examples of records include audit reports, requests for change, incident reports, individual training records and invoices sent to customers.	NOTE 2 Examples of records include audit reports, requests for change, incident reports, individual training records and invoices sent to customers.	✔					No change, left in for context of other changes.
2.10 release	**2.10 release**	✔					No change, left in for context of other changes.
2.11 request for change	**2.11 request for change**	✔					No change, left in for context of other changes.
2.12 service desk	**2.12 service desk**	✔					No change, left in for context of other changes.
2.13 service level agreement (SLA)	**2.13 service level agreement (SLA)**	✔					No change, left in for context of other changes.

Table 6 – The Part 1 differences (continued)

BS 15000-1	ISO/IEC 20000-1	N\C	Form	Edit	Rec.	Req.	Commentary
2.14 service management	**2.14 service management**	✔					No change, left in for context of other changes.
	2.15 service provider			✔			Additional definition, but no difference to the requirements or recommendations. The inclusion of 'service provider' in clause 2 is linked to the standardization on this term throughout the standard.
	the organization aiming to achieve ISO/IEC 20000			✔			
3 Requirements for a management system	**3 Requirements for a management system**	✔					No change, left in for context of other changes.
3.1 Management responsibility	**3.1 Management responsibility**	✔					No change, left in for context of other changes.
Through leadership and actions, **top management** shall provide evidence of its commitment to developing, implementing and improving its service management capability within the context of the organization's business and customers' requirements.	Through leadership and actions, **top/executive management** shall provide evidence of its commitment to developing, implementing and improving its service management capability within the context of the organization's business and customers' requirements.			✔			Top management changed to top/executive management, to aid understanding of the requirement, particularly for an international audience. This is not a change to the original intent of this requirement.
3.2 Documentation requirements	**3.2 Documentation requirements**	✔					No change, left in for context of other changes.

Table 6 – The Part 1 differences (continued)

BS 15000-1	ISO/IEC 20000-1	N/C	Form	Edit	Rec.	Req.	Commentary
Service providers shall provide **documentation** and records to ensure effective planning, operation and control of service management.	Service providers shall provide **documents** and records to ensure effective planning, operation and control of service management.			✔			Correction to the original wording in BS 15000, in which '*documentation*' was used instead of '*document*'. This is clarification of the intent of this clause, not a change. Documentation in the subclause heading is used for both documents and records.
Procedures and responsibilities shall be established for the creation, review, approval, **maintenance and control** of the various types of documents and records.	Procedures and responsibilities shall be established for the creation, review, approval, **maintenance, disposal and control** of the various types of documents and records.					✔	Addition of new requirement for the disposal of documents and records. This is a new requirement.
3.3 Competence, awareness and training	**3.3 Competence, awareness and training**	✔					No change, left in for context of other changes.
4 Planning and implementing service management	**4 Planning and implementing service management**	✔					No change, left in for context of other changes.
NOTE See also Introduction.			✔				Cross-reference back to the Introduction is no longer required.

Table 6 – The Part 1 differences (continued)

BS 15000-1	ISO/IEC 20000-1	N\C	Form	Edit	Rec.	Req.	Commentary
-	NOTE The methodology known as "Plan-Do-Check-Act" (PDCA) can be applied to all processes. PDCA can be described as follows: a) **Plan**: establish the objectives and processes necessary to deliver results in accordance with customer requirements and the organization's policies; b) **Do**: implement the processes; c) **Check**: monitor and measure processes and services against policies, objectives and requirements and report the results; d) **Act**: take actions to continually improve process performance.		✔				Moved from Introduction, wording unchanged. Provides an explanation of the requirements in clauses 4.1 to 4.4. No change to requirements or recommendations as this is a NOTE, which is advice, recommendations or explanation even when used in a requirements clause in Part 1 / specification. Bullets converted to alpha, from numeric.
	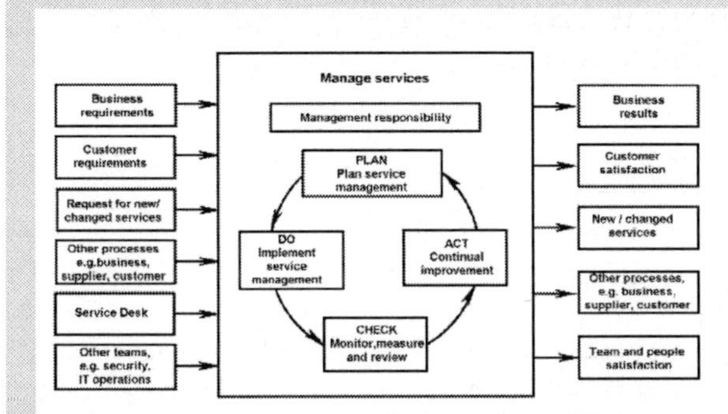		✔				Moved from Introduction, figure unchanged. Text and Figure provide an explanation of the requirements in clauses 4.1 to 4.4. No change to requirements or recommendations
	Figure 2 — Plan-Do-Check-Act methodology for service management processes		✔				Moved from Introduction, no change to wording.

Table 6 – The Part 1 differences (continued)

BS 15000-1	ISO/IEC 20000-1	N\C	Form	Edit	Rec.	Req.	Commentary
	The model shown in Figure 2 illustrates the process and process linkages presented in clauses **4 to 10.**		✔				Change to cross-reference.
4.1 Plan service management (Plan)	**4.1 Plan service management (Plan)**	✔					No change, left in for context of other changes.
Service management shall be planned.	Service management shall be planned.	✔					No change, left in for context of other changes.
The plans shall at a minimum define:	The plans shall at a minimum define:	✔					No change, left in for context of other changes.
a) the scope of **service management within the organization;**	a) the scope of the **service provider's service management;**			✔			Clarification from standardization of a term.
b) the objectives and requirements that are to be achieved by service management;	b) the objectives and requirements that are to be achieved by service management;	✔					No change, left in for context of other changes.
c) the processes that are to be executed;	c) the processes that are to be executed;	✔					No change, left in for context of other changes.
d) the framework of management roles and responsibilities, including **the process owner** and …	d) the framework of management roles and responsibilities, including the **senior responsible owner, process owner…**.					✔	Addition of the senior responsible owner to the minimum requirements for the plan.
…management of **third party suppliers;**	…and management of **suppliers;**			✔			Clarification from standardization of a term.
e) the interfaces between service management processes and the manner in which the activities are to be co-ordinated;	e) the interfaces between service management processes and the manner in which the activities are to be co-ordinated;	✔					No change, left in for context of other changes.
f) the approach to be taken in identifying, assessing and managing issues and risks to the achievement of the defined objectives;	f) the approach to be taken in identifying, assessing and managing issues and risks to the achievement of the defined objectives;	✔					

Table 6 – The Part 1 differences (continued)

BS 15000-1	ISO/IEC 20000-1	N\C	Form	Edit	Rec.	Req.	Commentary
g) the approach for interfacing to projects that are creating or modifying services;	g) the approach for interfacing to projects that are creating or modifying services;	✔					
h) the resources, facilities and budget necessary to achieve the defined objectives;	h) the resources, facilities and budget necessary to achieve the defined objectives;	✔					
i) tools as appropriate to support the processes; and	i) tools as appropriate to support the processes; and	✔					
j) how the quality of the service will be managed, audited and improved.	j) how the quality of the service will be managed, audited and improved.	✔					
4.2 Implement service management and provide the services (Do)	**4.2 Implement service management and provide the services (Do)**	✔					No change, left in for context of other changes.
The **organization** shall implement the service management plan to manage and deliver the services, including:	The **service provider** shall implement the service management plan to manage and deliver the services, including:			✔			Clarification from the use of a standard term.
a) allocation of funds and budgets;	a) allocation of funds and budgets;	✔					
b) allocation of roles and responsibilities;	b) allocation of roles and responsibilities;	✔					
c) documenting and maintaining the policies, plans, procedures and definitions for each process or set of processes;	c) documenting and maintaining the policies, plans, procedures and definitions for each process or set of processes;	✔					
d) identification and management of risks to the service;	d) identification and management of risks to the service;	✔					No change, left in for context of other changes.
e) managing teams, e.g. recruiting and developing appropriate staff and managing staff continuity;	e) managing teams, e.g. recruiting and developing appropriate staff and managing staff continuity;	✔					
f) managing facilities and budget;	f) managing facilities and budget;	✔					
g) managing the teams including service desk and operations;	g) managing the teams including service desk and operations;	✔					

Table 6 – The Part 1 differences (continued)

BS 15000-1	ISO/IEC 20000-1	N/C	Form	Edit	Rec.	Req.	Commentary
h) reporting progress against the plans; and	h) reporting progress against the plans; and	✔					
i) co-ordination of service management processes.	i) co-ordination of service management processes.	✔					
4.3 Monitoring, measuring and reviewing (Check)	**4.3 Monitoring, measuring and reviewing (Check)**	✔					
The **organization** shall apply suitable methods for monitoring and, where applicable, measurement of the service management processes.	The **service provider** shall apply suitable methods for monitoring and, where applicable, measurement of the service management processes.			✔			Clarification from the use of a standard term.
An audit programme shall be planned, taking into consideration the status and importance of the processes and areas to be audited, as well as the results of previous audits.	An audit programme shall be planned, taking into consideration the status and importance of the processes and areas to be audited, as well as the results of previous audits.	✔					No change, left in for context of other changes.
The audit criteria, scope, frequency and methods **shall be defined.**	The audit criteria, scope, frequency and methods **shall be defined in a procedure.**					✔	Minor change to the requirement from identification of the specific nature of how information will be documented.
4.4 Continuous improvement (Act)	**4.4 Continual improvement (Act)**			✔			This is correction of the word '*continuous*' by use of '*continual*' in line with other management system standards. This is not a change to the intentions of this clause and is therefore not coded as a change to requirements.
	4.4.1 Policy		✔				New subclause heading for ISO formatting.
There shall be a published policy on service improvement.	There shall be a published policy on service improvement.	✔					No change, left in for context of other changes.

Table 6 – The Part 1 differences (continued)

BS 15000-1	ISO/IEC 20000-1	N\C	Form	Edit	Rec.	Req.	Commentary
Any non-compliance with the standard or the service management plans shall be remedied.	Any non-compliance with the standard or the service management plans shall be remedied.	✔					
Roles and responsibilities for service improvement activities shall be clearly defined.	Roles and responsibilities for service improvement activities shall be clearly defined.	✔					
	4.4.2 Management of improvements		✔				New subclause heading for ISO formatting.
All suggested service improvements shall be assessed, recorded, prioritized and authorized.	All suggested service improvements shall be assessed, recorded, prioritized and authorized.	✔					
A service improvement plan shall be used to control the activity.	**A plan** shall be used to control the activity.			✔			Clarification from standardization of a term.
The **organization** shall have a process in place to identify, measure, report and manage improvement activities on an ongoing basis.	The **service provider** shall have a process in place to identify, measure, report and manage improvement activities on an ongoing basis.			✔			Clarification from standardization of a term.
This shall include:	This shall include:	✔					No change, left in for context of other changes.
a) improvements to an individual process that can be implemented by the process owner with the usual staff resources, e.g. performing individual corrective and preventive actions; and	a) improvements to an individual process that can be implemented by the process owner with the usual staff resources, e.g. performing individual corrective and preventive actions; and	✔					
b) improvements across the organization or across more than one process.	b) improvements across the organization or across more than one process.	✔					
	4.4.3 Activities		✔				New subclause heading for ISO formatting.
The **organization** shall perform activities to:	The **service provider** shall perform activities to:			✔			Clarification from standardization of a term.

Table 6 – The Part 1 differences (continued)

BS 15000-1	ISO/IEC 20000-1	N/C	Form	Edit	Rec.	Req.	Commentary
1) collect and analyse data to baseline and benchmark the **organization's** capability to…….	a) collect and analyse data to baseline and benchmark the **service provider's** capability to ……..			✔			Clarification from standardization of a term.
……. manage and deliver **service management;**	………manage and deliver **service and service management processes;**					✔	Change to requirements, extending the scope of baselining and benchmarking.
2) identify, plan and implement improvements;	b) identify, plan and implement improvements;		✔				Minor change to formatting, numeric bullets converted to alpha.
3) consult with all parties involved;	c) consult with all parties involved;		✔				
4) set targets for improvements in quality, costs and resource utilization;	d) set targets for improvements in quality, costs and resource utilization;		✔				
5) consider relevant inputs about improvements from all the service management processes;	e) consider relevant inputs about improvements from all the service management processes;		✔				
6) measure, report and communicate the service improvements;	f) measure, report and communicate the service improvements;		✔				
7) revise the service management **policies, plans and procedures** where necessary; and	g) revise the service management **policies, processes, procedures and plans** where necessary; and					✔	Change to requirement from inclusion of '*processes*'.
8) ensure that all approved actions are delivered and that they achieve their intended objectives.	h) ensure that all approved actions are delivered and that they achieve their intended objectives.		✔				Minor change to formatting, numeric bullets converted to alpha.

Table 6 – The Part 1 differences (continued)

BS 15000-1	ISO/IEC 20000-1	N\C	Form	Edit	Rec.	Req.	Commentary
Major service improvements shall be managed as a project or several projects.				✔			Clause 4.4 is on planning and service improvements so this requirement was superfluous and has been deleted. As it was superfluous deletion has not changed the requirements, although deletion may affect audit, training and consulting material based on BS 15000.
5 Planning and implementing new or changed services	**5 Planning and implementing new or c hanged services**	✔					No change, left in for context of other changes.
*Objective: To ensure that new services and changes to services will be deliverable and manageable at the **right** cost and service quality.*	*Objective: To ensure that new services and changes to services will be deliverable and manageable at the **agreed** cost and service quality.*				✔		Agreed cost instead of right cost, clarification and consistency of objective across Parts 1 and 2. Objectives are not part of the requirements and so this is a change to advice.
The plans shall include:	The plans shall include:	✔					No change, left in for context of other changes.
a) the roles and responsibilities for implementing, operating and maintaining the new or changed service including activities to be performed by customers **and third party suppliers;**	a) the roles and responsibilities for implementing, operating and maintaining the new or changed service including activities to be performed by customers **and suppliers;**			✔			Clarification from standardization of a term.
b) changes to the existing service management framework and services;	b) changes to the existing service management framework and services;	✔					No change, left in for context of other changes.
c) communication to the relevant parties;	c) communication to the relevant parties;	✔					

Table 6 – The Part 1 differences (continued)

BS 15000-1	ISO/IEC 20000-1	N/C	Form	Edit	Rec.	Req.	Commentary
d) new or changed contracts and agreements to align with the changes in business need;	d) new or changed contracts and agreements to align with the changes in business need;	✔					
e) manpower and recruitment requirements;	e) manpower and recruitment requirements;	✔					
f) skills and training requirements, **<u>e.g. end users,</u>** technical support;	f) skills and training requirements, <u>e.g. users,</u> technical support;			✔			Clarification from standardization of a term.
g) processes, measures, methods and tools to be used in connection with the new or changed service, e.g. capacity management, financial management;	g) processes, measures, methods and tools to be used in connection with the new or changed service, e.g. capacity management, financial management;	✔					No change, left in for context of other changes.
h) budgets and time-scales;	h) budgets and time-scales;	✔					
i) service acceptance criteria; and	i) service acceptance criteria; and	✔					
j) the expected outcomes from operating the new service expressed in measurable terms.	j) the expected outcomes from operating the new service expressed in measurable terms.	✔					
The service provider shall report on the outcomes achieved by the new or changed service against those planned following its implementation.	The service provider shall report on the outcomes achieved by the new or changed service against those planned following its implementation.	✔					
A post implementation review comparing actual outcomes against those planned, shall be performed through **<u>the change management process and the results reported to the relevant parties.</u>**	A post implementation review comparing actual outcomes against those planned shall be performed through **<u>the change management process.</u>**			✔			Clarification and not a change to the requirement, as reporting results to the relevant parties is inherent in the change management process and does not need to be explicitly included.

Table 6 – The Part 1 differences (continued)

BS 15000-1	ISO/IEC 20000-1	N\C	Form	Edit	Rec.	Req.	Commentary
6 Service delivery process	**6 Service delivery process**	✔					No change, left in for context of other changes.
6.1 Service level management	**6.1 Service level management**	✔					No change, left in for context of other changes.
SLAs, together with supporting service agreements, **third party contracts** and corresponding procedures, shall be agreed by all relevant parties and recorded.	SLAs, together with supporting service agreements, **supplier contracts** and corresponding procedures, shall be agreed by all relevant parties and recorded.			✔			Clarification from standardization of a term.
The SLAs shall be under **change control.**	The SLAs shall be under **the control of the change management process.**			✔			Clarification and not a change to the requirement or the original intent of this clause.
The SLAs shall be maintained by regular reviews by the parties to ensure that they are up to date and remain effective over time	The SLAs shall be maintained by regular reviews by the parties to ensure that they are up to date and remain effective over time	✔					No change, left in for context of other changes.
Service levels shall be monitored and reported against targets, showing both current and trend information.	Service levels shall be monitored and reported against targets, showing both current and trend information.	✔					No change, left in for context of other changes.
The reasons for non-conformance shall be reported and reviewed …..	The reasons for non-conformance shall be reported and reviewed.		✔				Change to punctuation arising from the original sentence being split into two sentences as part of the clarification of the requirement.
…..and shall provide input to the service improvement plan.	**Actions for improvement identified during this process shall be recorded and provide input into a plan for improving the service.**			✔			Clarification and not a change to the requirement.

Table 6 – The Part 1 differences (continued)

BS 15000-1	ISO/IEC 20000-1	N/C	Form	Edit	Rec.	Req.	Commentary
6.2 Service reporting	**6.2 Service reporting**	✔					No change, left in for context of other changes.
Service reports shall be produced to meet identified needs and customer requirements.	Service reports shall be produced to meet identified needs and customer requirements.	✔					No change, left in for context of other changes.
<u>Typically service reporting shall include</u>:	<u>Service reporting shall include:</u>					✔	Change to requirement as service reporting must now always include items a) to f), due to the deletion of 'typical'.
<u>a) performance against service level targets;</u>	<u>a) performance against service level targets;</u>					✔	
<u>b) non-compliance and issues, e.g. against the SLA, security breech;</u>	<u>b) non-compliance and issues, e.g. against the SLA, security breech;</u>					✔	
<u>c) workload characteristics, e.g. volume, resource utilisation;</u>	<u>c) workload characteristics, e.g. volume, resource utilization;</u>					✔	
<u>d) performance reporting following major events, e.g. major incidents and changes;</u>	<u>d) performance reporting following major events, e.g. major incidents and changes;</u>					✔	
<u>e) trend information;</u>	<u>e) trend information;</u>					✔	
<u>f) satisfaction analysis.</u>	<u>f) satisfaction analysis.</u>					✔	
<u>6.3 Availability and service continuity management</u>	**<u>6.3 Service continuity and availability management</u>**			✔			Standardization of the clause heading across Parts 1 and 2.
*Objective: To ensure that agreed **<u>obligations to customers can be met in all circumstances.</u>***	*Objective: To ensure that agreed **<u>service continuity and availability commitments to customers can be met in all circumstances.</u>***				✔		Change to the objectives to clarify which obligations are referred to. As objectives are not part of the requirements this is a change to the recommendations.

Table 6 – The Part 1 differences (continued)

BS 15000-1	ISO/IEC 20000-1	N\C	Form	Edit	Rec.	Req.	Commentary
Availability and service continuity plans shall be developed and reviewed at least annually to ensure that requirements are met as agreed in all circumstances from normal through to a major loss of service.	Availability and service continuity plans shall be developed and reviewed at least annually to ensure that requirements are met as agreed in all circumstances from normal through to a major loss of service.	✔					No change, left in for context of other changes.
These plans shall be maintained to ensure that they reflect agreed changes required by the business.	These plans shall be maintained to ensure that they reflect agreed changes required by the business.	✔					No change, left in for context of other changes.
	The availability and service continuity plans shall be re-tested at every major change to the business environment.					✔	Additional requirement.
The change management process shall assess the impact of any change on the availability and service continuity plan.	The change management process shall assess the impact of any change on the availability and service continuity plan.	✔					
Availability shall be measured and recorded.	Availability shall be measured and recorded.	✔					No change, left in for context of other changes
Unplanned non-availability shall be investigated and appropriate actions taken.	Unplanned non-availability shall be investigated and appropriate actions taken.	✔					
Where possible, potential issues shall be predicted and preventive action taken.	**NOTE Where possible, potential issues should be predicted and preventive action taken.**					✔	A requirement changed to a recommendation by conversion to a NOTE. 'Where possible...' is not compatible with a requirement which must be met, so this clause was open to misinterpretation.

Table 6 – The Part 1 differences (continued)

BS 15000-1	ISO/IEC 20000-1	N/C	Form	Edit	Rec.	Req.	Commentary
6.4 Budgeting and accounting for IT services	**6.4 Budgeting and accounting for IT services**	✔					No change, left in for context of other changes.
NOTE This section covers budgeting and accounting for IT services. In practice, many **organizations** will be involved in charging for such services.	NOTE This section covers budgeting and accounting for IT services. In practice, many **service providers** will be involved in charging for such services.			✔			Clarification from standardization of a term.
However, since charging is an optional activity, it is not covered by the standard.	However, since charging is an optional activity, it is not covered by the standard.	✔					No change, left in for context of other changes.
Organizations are recommended that where charging is in use, the mechanism for doing so is fully defined and understood by all parties.	**Service providers** are recommended that where charging is in use, the mechanism for doing so is fully defined and understood by all parties.			✔			Clarification from standardization of a term.
All accounting practices in use should be aligned to the wider accountancy practices of the **organisation**.	All accounting practices in use should be aligned to the wider accountancy practices of the **service provider's organization**.			✔			Clarification from standardization of a term.
There shall be clear policies **and procedures** for:	There shall be clear policies **and processes for**:					✔	Change to requirements, procedure change to process.
a) budgeting, and accounting for all components including IT assets, shared resources, overheads, **third-party supplied service**, people, insurance and licences;	a) budgeting, and accounting for all components including IT assets, shared resources, overheads, **externally supplied service**, people, insurance and licences;			✔			Clarification from standardization of a term.
b) **apportioning and allocating all indirect costs to relevant services;**	b) **apportioning indirect costs and allocating direct costs to services;**			✔			Correction of the requirement (indirect costs are apportioned, direct costs allocated). This is not a change to the intent of the requirement. Clarification from standardization of a term.

Table 6 – The Part 1 differences (continued)

BS 15000-1	ISO/IEC 20000-1	N\C	Form	Edit	Rec.	Req.	Commentary
c) effective financial control and authorization.	c) effective financial control and authorization.	✔					No change, left in for context of other changes.
6.5 Capacity management	**6.5 Capacity management**	✔					No change, left in for context of other changes.
*Objective: To ensure that the **organization** has, at all times, sufficient capacity to **meet the current and future agreed demands of the business.***	*Objective: To ensure that the **service provider** has, at all times, sufficient capacity **to meet the current and future agreed demands of the customer's business needs.***				✔		Change to the objectives, as objectives are not part of the requirements this is a change to the recommendations.
6.6 Information security management	**6.6 Information security management**	✔					No change, left in for context of other changes.
NOTE **BS ISO/IEC 17799 (BS 7799-1)** provides guidance on information security management.	NOTE **ISO/IEC 17799, Information technology — Security techniques — Code of practice for information security management** provides guidance on information security management.				✔		As this is a change to a NOTE it is equivalent to a change to a recommendation even though this is a Part 1 / specification. Changes to references to BS 7799 have also been made to ISO/IEC 20000.
Implementation of the requirements in this specification may not satisfy all the requirements that are necessary to obtain certification against BS 7799-2.					✔		As this is also part of the NOTE in BS 15000, deletion does not change the requirements. This provides advice, but refers to a national standard which has not been superseded by an ISO/IEC standard.

Table 6 – The Part 1 differences (continued)

BS 15000-1	ISO/IEC 20000-1	N/C	Form	Edit	Rec.	Req.	Commentary
Organizations certified to BS 7799 will satisfy the security requirements within BS 15000- 1.					✔		As this is also part of the NOTE in BS 15000, deletion does not change the requirements. This provides advice, but refers to a national standard which has not been superseded by an ISO/IEC standard.
Arrangements that involve **third-party access to** information systems and services shall be based on a formal agreement that defines all necessary security requirements.	Arrangements that involve **external organizations having** access to information systems and services shall be based on a formal agreement that defines all necessary security requirements.			✔			Clarification from standardization of a term.
Mechanisms shall be in place to enable the types, volumes and impacts of security incidents and malfunctions to be quantified and monitored.....	Mechanisms shall be in place to enable the types, volumes and impacts of security incidents and malfunctions to be quantified and monitored.	✔					No change, left in for context of other changes.
....**and to provide input to the service improvement plan.**	**Actions for improvements identified during this process shall be recorded and provide input into a plan for improving the service.**			✔			Clarification from standardization of a term.
7 Relationship processes	**7 Relationship processes**	✔					No change, left in for context of other changes.
	7.1 General		✔				New clause heading as part of ISO formatting.
	Relationship processes describe the two related aspects of Supplier Management and Business Relationship Management.		✔				New text as part of ISO formatting.
7.1 Business relationship management	**7.2 Business relationship management**		✔				Clause heading re-numbered, both parts aligned.

Table 6 – The Part 1 differences (continued)

BS 15000-1	ISO/IEC 20000-1	N\C	Form	Edit	Rec.	Req.	Commentary
There shall be a **complaint procedure.**	There shall be a **complaints process**.					✔	Procedure changed to process, changing the requirement.
The service provider shall have a named individual or individuals who are responsible for managing customer satisfaction and the whole business relationship process.	The service provider shall have a named individual or individuals who are responsible for managing customer satisfaction and the whole business relationship process.	✔					No change, left in for context of other changes.
A process shall exist for obtaining and acting upon feedback from regular customer satisfaction measurements.	A process shall exist for obtaining and acting upon feedback from regular customer satisfaction measurements.	✔					No change, left in for context of other changes.
Actions for improvement identified during this process shall be recorded and **input into the service improvement plan**.	Actions for improvement identified during this process shall be recorded and **input into a plan for improving the service.**			✔			Clarification from standardization of a term.
7.2 Supplier management	**7.3 Supplier management**		✔				Clause heading re-numbered.
Objective: To manage ***third party suppliers*** *to ensure the provision of seamless, quality services.*	*Objective: To manage* ***suppliers*** *to ensure the provision of seamless, quality services.*			✔			Clarification from standardization of a term. This does not change the requirements as the objectives are not part of the requirements.
NOTE 1 The scope of this standard excludes the procurement of the suppliers.	NOTE 1 The scope of this standard excludes the procurement of the suppliers.	✔					No change, left in for context of other changes.
NOTE 2 **Third party suppliers** may be used by the service provider for provision of some part of the service. It is the service provider who needs to demonstrate conformity to these supplier management processes. Complex relationships may be present as demonstrated in the diagram below which is used as an example:	NOTE 2 **Suppliers** may be used by the service provider for provision of some part of the service. It is the service provider who needs to demonstrate conformity to these supplier management processes. Complex relationships may be present as demonstrated in the diagram below which is used as an example:			✔			Clarification from standardization of a term.

Table 6 – The Part 1 differences (continued)

BS 15000-1	ISO/IEC 20000-1	N/C	Form	Edit	Rec.	Req.	Commentary
A formal process shall exist to deal with contractual disputes.	**A process** shall exist to deal with contractual disputes.			✔			In the context of the standard all processes in the scope are formal, so the use of formal process added no value and formal has been deleted. This is not a change to requirements.
Performance against service level targets shall be monitored and reviewed.	Performance against service level targets shall be monitored and reviewed.	✔					No change, left in for context of other changes.
Actions for improvement identified during this process shall be recorded **and input into the service improvement plan.**	Actions for improvement identified during this process shall be recorded **and input into a plan for improving the service.**			✔			Clarification from standardization of a term.
8 Resolution processes	**8 Resolution processes**	✔					No change, left in for context of other changes.
	8.1 Background		✔				New clause heading as part of formatting.
	Incident and problem management are separate processes, although they are closely linked.		✔				New text, aligning the structure of Parts 1 and 2.
8.1 Incident management	**8.2 Incident management**		✔				Change to clause heading number, aligning Parts 1 and 2.
Procedures shall be adopted to manage the impact **of service incidents.**	Procedures shall be adopted to manage the impact **of incidents.**			✔			The word service was deleted as it added no value. This is not a change to requirements.
Major incidents shall be classified and managed according to **a defined process.**	Major incidents shall be classified and managed according to **a process.**			✔			The word defined was deleted as it added no value. This is not a change to the requirements as all processes are defined and formal in the context of ISO/IEC 20000.

Table 6 – The Part 1 differences (continued)

BS 15000-1	ISO/IEC 20000-1	N\C	Form	Edit	Rec.	Req.	Commentary
8.2 Problem management	**8.3 Problem management**			✔			Change to clause heading number, aligning Parts 1 and 2.
*Objective: To minimize disruption to the business by proactive identification and analysis of the cause **of service incidents** and by managing problems to closure.*	*Objective: To minimize disruption to the business by proactive identification and analysis of the cause **of incidents** and by managing problems to closure.*			✔			The word service was deleted as it added no value. This is not a change to requirements and Objectives provide only advice/recommendations.
Preventive **action will be** taken to reduce potential problems, e.g. following trend analysis of incident volumes and types.	Preventive **action shall be** taken to reduce potential problems, e.g. following trend analysis of incident volumes and types.	✔					Correction of an error in BS 15000. This does not change the original intent of this clause, which was for this requirement to be compulsory.
Actions for improvement identified during this process shall be recorded and input into **the service improvement plan.**	Actions for improvement identified during this process shall be recorded a**nd input into a plan for improving the service.**	✔					Clarification from standardization of a term.
9 Control processes	**9 Control processes**		✔				No change, left in for context of other changes.
9.1 Configuration management	**9.1 Configuration management**		✔				No change, left in for context of other changes.
NOTE 1 Financial asset accounting falls outside the scope of this section.	*Objective: To define and control the components of the service and infrastructure and maintain accurate configuration information.*		✔				Re-ordering of the text so that the Objective is after the subclause title.
NOTE 2 BS EN ISO 10007 provides further guidelines on configuration management.			✔				NOTE (which is not part of the requirements) has been dropped. A reference to the latest version of this standard is included in the Bibliography.

Table 6 – The Part 1 differences (continued)

BS 15000-1	ISO/IEC 20000-1	N/C	Form	Edit	Rec.	Req.	Commentary
	There shall be an integrated approach to change and configuration management planning.		✔				Re-ordering of text, no change to requirements.
Objective: To define and control the components of the service and infrastructure and maintain accurate configuration information.	The service provider shall define the interface to financial asset accounting processes.		✔				Re-ordering of the text so that 'Objective' is immediately after the subclause title.
There shall be an integrated approach to change and configuration management planning.	NOTE Financial asset accounting falls outside the scope of this section.		✔				Re-ordering of the text so that the NOTE is after the Objective and not before, as it was in BS 15000-1.
The service provider shall define the interface to financial asset accounting processes.			✔				Re-ordering of text, no change to requirements.
Configuration management shall provide the mechanisms for identifying, controlling and tracking versions of identifiable components of the service and infrastructure.	Configuration management shall provide the mechanisms for identifying, controlling and tracking versions of identifiable components of the service and infrastructure.	✔					No change, left in for context of other changes.
The degree of control shall be sufficient to meet the business needs, risk of failure and service criticality.	**It shall be ensured that the degree of control is sufficient to meet the business needs, risk of failure and service criticality.**			✔			Clarification from re-wording. This is not a change of the original intent of this requirement.
Master copies of digital configuration items shall be controlled in secure physical or electronic libraries and **reference to the** configuration records, e.g. software, testing products, support documents.	Master copies of digital configuration items shall be controlled in secure physical or electronic libraries and **referenced to the** configuration records, e.g. software, testing products, support documents.		✔				Correction of minor typing error in BS 15000-1.
All configuration items shall be uniquely identifiable and recorded in a CMDB to which update access shall be strictly controlled.	All configuration items shall be uniquely identifiable and recorded in a CMDB to which update access shall be strictly controlled.	✔					No change, left in for context of other changes.

Table 6 – The Part 1 differences (continued)

BS 15000-1	ISO/IEC 20000-1	N\C	Form	Edit	Rec.	Req.	Commentary
The CMDB shall be actively managed and verified to ensure its reliability and accuracy.	The CMDB shall be actively managed and verified to ensure its reliability and accuracy.	✔					No change, left in for context of other changes.
The status **to** configuration items, their versions, location, related changes and problems and associated documentation shall be visible to those who require it.	The status **of** configuration items, their versions, location, related changes and problems and associated documentation shall be visible to those who require it.		✔				Correction of minor error in BS 15000-1.
Configuration audit procedures shall include recording deficiencies **and instigating** corrective actions and reporting on the outcome.	Configuration audit procedures shall include recording deficiencies**, initiating** corrective actions and reporting on the outcome.			✔			Change to assist interpretation of the requirement by the international users of this standard. This is not a change to the original intent of BS 15000-1.
9.2 Change management	**9.2 Change management**	✔					No change, left in for context of other changes.
The scheduled implementation dates of changes shall be used as the basis for change and release scheduling.	The scheduled implementation dates of changes shall be used as the basis for change and release scheduling.	✔					No change, left in for context of other changes.
A forward schedule of change shall be maintained and communicated to relevant parties.	**A schedule that contains details of all the changes approved for implementation and their proposed implementation dates shall be maintained and communicated to relevant parties.**			✔			Clarification of the requirement, not a change.
Change records shall be analysed regularly to detect increasing levels of changes, frequently recurring types, emerging trends and other relevant information.	Change records shall be analysed regularly to detect increasing levels of changes, frequently recurring types, emerging trends and other relevant information.	✔					No change, left in for context of other changes.
The results and conclusions drawn from change analysis shall be recorded.	The results and conclusions drawn from change analysis shall be recorded.	✔					No change, left in for context of other changes.

Table 6 – The Part 1 differences (continued)

BS 15000-1	ISO/IEC 20000-1	N/C	Form	Edit	Rec.	Req.	Commentary
Actions for improvement identified from change management shall be recorded and **input into the service improvement plan.**	Actions for improvement identified from change management shall be recorded and **input into a plan for improving the service.**			✔			Clarification from standardization of a term.
10 Release process	**10 Release process**	✔	✔				No change, left in for context of other changes.
10.1 Release management	**10.1 Release management process**			✔			Change to heading tile to make the structure of both Pats 1 and 2 consistent.
	NOTE The release management process should be integrated with the configuration and change management processes.				✔		New NOTE giving a recommendation / advice, not a requirement. The requirements of ISO/IEC 20000 can only be met if processes are suitably integrated, so this NOTE is emphasis of a particularly important aspect of process integration. The recommendation given in this NOTE was originally in BS 15000-2.
The process shall include the manner in which the release shall be **backed-out** or remedied if unsuccessful.	The process shall include the manner in which the release shall be **reversed** or remedied if unsuccessful.			✔			Clarification from standardization of a term.
Plans shall record the release dates and deliverables and refer to related change requests, known errors and problems.	Plans shall record the release dates and deliverables and refer to related change requests, known errors and problems.	✔					No change, left in for context of other changes.

Table 6 – The Part 1 differences (continued)

BS 15000-1	ISO/IEC 20000-1	N\C	Form	Edit	Rec.	Req.	Commentary
They shall be communicated to incident management.	**The release management process shall pass suitable information to the incident management process.**			✔			Clarification from changes to wording, avoiding any confusion between processes and functional groups with similar names to the processes they perform. This clarifies the requirement and emphasises the process basis of ISO/IEC 20000.
Success and failure of releases shall be measured.	Success and failure of releases shall be measured.	✔					No change, left in for context of other changes.
Measurements shall include incidents related to a release in the period following a release.	Measurements shall include incidents related to a release in the period following a release.	✔					No change, left in for context of other changes.
Analysis shall include assessment of the impact on the business, IT operations and support staff resources, and shall provide input to **the service improvement plan.**	Analysis shall include assessment of the impact on the business, IT operations and support staff resources, and shall provide input to **a plan for improving the service.**			✔			Clarification from standardization of a term.
Bibliography	**Bibliography**	✔					
Standards publications				✔			Original contents of the BS 15000 Bibliography are not allowed under ISO/IEC formatting rules.
	[1] ISO/IEC 20000, *Information technology — service management—Part 2: Code of Practice*			✔			Reference corrected to reflect ISO/IEC status.
	Plus 15 other ISO or ISO/IEC standards Reference corrected to reflect ISO/IEC status.			✔			

Table 6 – The Part 1 differences (continued)

BS 15000-1	ISO/IEC 20000-1	N/C	Form	Edit	Rec.	Req.	Commentary
Non-standards publications			✔				Out of date or not allowed under formatting rules.
	National annex NA (informative)		✔				Additional reference only in BS ISO/IEC 20000.
	National bibliography		✔				Additional reference only in BS ISO/IEC 20000.

CHAPTER 7

Detailed changes to Part 2

Introduction

This chapter provides details of the changes to the code of practice, ISO/IEC 20000-2. It supplements the details in Chapters 3 and 4. There are many relatively minor changes to ISO/IEC 20000-2. Most of the changes are formatting or clause numbering changes, including the introduction of subclauses to allow, and simplify, cross-referencing of requirements. Some structural changes have been made to align Parts 1 and 2 at process subclause level.

The column headings described in detail in Chapter 5 are used in Table 7 to distinguish between the different types of changes to Part 2. Unchanged text is not shown in Table 7, unless it is necessary to give context to requirements or recommendations that have changed.

Text that has been changed is normally underlined. Where whole sentences or a whole paragraph of text are new or have been completely deleted there is only text in one of the columns. This makes the change easy to see, and is therefore not underlined. Additional headings or deleted headings are not underlined for the same reason.

To make it easier to identify a NOTE in Table 7, NOTEs are included as serif fonts, with the rest of the clause text as san serif font. The difference between a NOTE and normal text in Part 2 (the code of practice) is of much less significance than the difference between a NOTE in Part 1 (the specification).

Table 7 – The Part 2 differences

B 15000-2	ISO/IEC 2000-2	N/C	Form	Edit	Rec	Req	Commentary
IT Service Management – Part 2: Code of practice for service management	**Information technology – Service management – Part 2: Code of practice**			✔			New title in line with ISO/IEC practices.
	National Foreword		✔				
	This British Standard reproduces verbatim ISO/IEC 20000-2:2005 and implements it as the UK national standard.		✔				Standard text for a BS version of an ISO/IEC standard.
	It supersedes BS 15000-2:2002 which is withdrawn.		✔				Standard text for a BS version of an ISO/IEC standard.
			✔				The rest of the National Foreword is identical to the National Foreword in ISO/IEC 2000-1 and is not repeated here.
Foreword	**Foreword**	✔					No change, left in for context of other changes.
	See the Foreword in Part 1			✔			The Part 2 Foreword is very similar to that in ISO/IEC 20000, except for the reference to the other part of ISO/IEC 20000.
Introduction	**Introduction**	✔					No change, left in for context of other changes.
Why is part 2 needed?			✔				Heading dropped as superfluous.

Table 7 – The Part 2 differences (continued)

BS 15000-2	ISO/IEC 20000-2	N/C	Form	Edit	Rec	Req	Commentary
	As a code of practice, this **part of ISO/IEC 20000** takes the form of guidance and recommendations.		✔				Moved from Foreword and reworded to reflect the international status and ISO/IEC formatting rules.
	It should not be quoted as if it were a specification and particular care should be taken to ensure that claims of compliance are not misleading.		✔				
	This part of ISO/IEC 20000 should be used in conjunction with **ISO/IEC 20000-1**, the specification associated with this code of practice.		✔				
With the increasing dependencies in support services and the diverse range of technologies available, **service suppliers** can struggle to maintain high levels of customer service.	With the increasing dependencies in support services and the diverse range of technologies available, **service providers** can struggle to maintain high levels of customer service.			✔			Clarification from standardization of a term.
Figure 1 — Service management processes			✔				Moved to the Scope.
1 Scope	**1 Scope**	✔					No change, left in for context of other changes.
BS 15000-2 represents an industry consensus on quality standards for IT service management processes.	**This part of ISO/IEC 20000** represents an industry consensus on quality standards for IT service management processes.		✔				Change in how the standard is referred to in line with ISO/IEC practices.
These service management processes deliver the best possible service to meet an **organization's business** needs within agreed resource levels, i.e. service that is professional, cost effective and with risks which are understood and managed.	These service management processes deliver the best possible service to meet a **customer's business needs** within agreed resource levels, i.e. service that is professional, cost effective and with risks which are understood and managed.			✔			Clarification from standardization of a term.
It also provides a framework for use by suppliers of service management tools.	It also provides a framework for use by suppliers of service management tools.	✔					No change, left in for context of other changes.

Table 7 – The Part 2 differences (continued)

BS 15000-2	ISO/IEC 20000-2	N/C	Form	Edit	Rec	Req	Commentary
	As a process based standard this code of practice is not intended for product assessment. However, organizations developing service management tools, products and systems may use both the specification and the code of practice to help them develop tools, products and systems that support best practice service management.			✔			Clarification and emphasis of the process basis of the standard; does not change the requirements or recommendations. Clarification of the process basis; does not change the requirements or recommendations. This wording was also added to ISO/IEC 20000-1.
BS 15000-2, provides guidance to auditors and offers assistance to **organizations** planning service improvements or to be audited against **BS 15000-1.**	**This part of ISO/IEC 20000** provides guidance to auditors and offers assistance to **service providers** planning service improvements or to be audited against **ISO/IEC 20000-1.**			✔			Change in how the standard is referred to in line with ISO/IEC practices. Clarification from standardization of a term.
	Figure 1 – Service management processes		✔				Figure moved from the Introduction, but not changed.
2 Terms and definitions	**2 Terms and definitions**	✔					
For the purposes of this part of **BS 15000** the terms and definitions given in **BS 15000-1** apply.	For the purposes of **this document** the terms and definitions given in **ISO/IEC 20000-1** apply.		✔				Change in how the standard is referred to in line with ISO/IEC practices.
3 The management system	**3 The management system**	✔					No change, left in for context of other changes.
3.1 Management responsibility	**3.1 Management responsibility**	✔					No change, left in for context of other changes.
The role of management in ensuring best practice processes are adopted and sustained is fundamental for any service provider to meet the requirements of **BS 15000-1.**	The role of management in ensuring best practice processes are adopted and sustained is fundamental for any service provider to meet the requirements of **ISO/IEC 20000-1.**		✔				

Table 7 – The Part 2 differences (continued)

BS 15000-2	ISO/IEC 20000-2	N/C	Form	Edit	Rec	Req	Commentary
The senior responsible owner's role should encompass resourcing for any **continuous** or project based service improvement activities.	The senior responsible owner's role should encompass resourcing for any **continual** or project based service improvement activities.			✔			The word continuous changed to continual. This is not a change to the original intent of the standard but it now aligns with other management system standards.
3.2 Documentation requirements	**3.2 Documentation requirements**	✔					No change, left in for context of other changes.
The following documents are normally considered suitable as evidence of service management planning:	The following documents are normally considered suitable as evidence of service management planning.	✔					No change, left in for context of other changes.
a) policies and plans;	a) policies and plans;	✔					
b) service documentation;	b) service documentation;	✔					
c) procedures;	c) procedures;	✔					
d) process;	d) processes;			✔			Correction of minor error in BS 15000, not a change to recommendations or requirements.
e) process control records.	e) process control records.	✔					
3.3 Managing documents			✔				Clause heading deleted, mapping Parts 1 and 2.
There should be a process for the creation and management of documents to help ensure that the characteristics described are met.	There should be a process for the creation and management of documents to help ensure that the characteristics described are met.	✔					No change, left in for context of other changes.
Documentation should be protected from damage due, for example, to poor environmental conditions and computer disasters.	Documentation should be protected from damage due, for example, to poor environmental conditions and computer disasters.	✔					No change, left in for context of other changes

Table 7 – The Part 2 differences (continued)

BS 15000-2	ISO/IEC 20000-2	N/C	Form	Edit	Rec	Req	Commentary
3.4 Competence, awareness and training	**3.3 Competence, awareness and training**		✔				Clause heading re-numbered.
Objective: To ensure that service management personnel are competent to undertaken their role.					✔		Objective deleted mapping Parts 1 and 2. Objective following Clause 3 heading applies to the whole of clause 3.
3.4.1 Introduction	***3.3.1 General***		✔				Subclause heading changed to align with ISO/IEC practices. Numbering changed to align Parts 1 and 2.
The **organization** should:	The **service provider** should:			✔			Clarification from standardization of a term.
a) determine the necessary competence for each role in service management;	a) determine the necessary competence for each role in service management;	✔					No change, left in to clarify where the new heading has been included.
b) ensure that personnel are aware of the relevance and importance of their activities within the wider business context and how they contribute to the achievement of quality objectives;	b) ensure that personnel are aware of the relevance and importance of their activities within the wider business context and how they contribute to the achievement of quality objectives;	✔					
c) maintain appropriate records of education, training, skills and experience;	c) maintain appropriate records of education, training, skills and experience;	✔					
d) provide training or take other action to satisfy these needs;	d) provide training or take other action to satisfy these needs;	✔					
e) evaluate the effectiveness of the actions taken.	e) evaluate the effectiveness of the actions taken.	✔					

Table 7 – The Part 2 differences (continued)

BS 15000-2	ISO/IEC 20000-2	N/C	Form	Edit	Rec	Req	Commentary
	3.3.2 Professional development		✔				New subclause heading for ISO formatting.
The service provider should develop and enhance the professional competence of their workforce.	The service provider should develop and enhance the professional competence of their workforce.	✔					No change, left in to clarify where the new heading has been included.
Among the measures taken to achieve this, the service provider should address the following:…..	Among the measures taken to achieve this, the service provider should address the following:……..		✔				Minor formatting change, bullet list has been changed from numeric to alpha.
Staff should be trained in the relevant aspects of service management **(e.g. via formal courses, self study, mentoring and on the job training)** and their team-working and leadership skills should be developed.	Staff should be trained in the relevant aspects of service management **(e.g. via training courses, self study, mentoring and on the job training)** and their team-working and leadership skills should be developed.			✔			Change to wording on recommended training, particularly for ease of understanding by an international audience.
	3.3.3 Approaches to be considered		✔				New subclause heading for ISO formatting.
In order to achieve teams of staff with **an appropriate** levels of competence the servic e provider should decide on the optimum mix of short term and permanent recruits.	In order to achieve teams of staff **with appropriate** levels of competence the service provider should decide on the optimum mix of short term and permanent recruits.		✔				Typo in original standard carried over, 'an' deleted.
The service provider should also decide on the optimum mix of new staff with the skills required and re-training of existing staff.	The service provider should also decide on the optimum mix of new staff with the skills required and re-training of existing staff.	✔					No change, left in for context of other changes.
NOTE The optimum balance of short term and permanent recruits is particularly important when the **supplier** is planning how to provide a service during and after major changes to the number and skills of the support staff.	NOTE The optimum balance of short term and permanent recruits is particularly important when the **service provider** is planning how to provide a service during and after major changes to the number and skills of the support staff.			✔			Clarification from standardization of a term. This is not a change from the original intent of the recommendation.

Table 7 – The Part 2 differences (continued)

BS 15000-2	ISO/IEC 20000-2	N/C	Form	Edit	Rec	Req	Commentary
Factors that should be considered when establishing the most suitable combination of approaches include:	Factors that should be considered when establishing the most suitable combination of approaches include:		✔				Minor formatting change to bullet list.
For all staff, the service provider should review each individual's performance at least annually and take appropriate action.	For all staff, the service provider should review each individual's performance at least annually and take appropriate action.	✔					No change, left in for context of other changes.
4 Planning and implementing service management	**4 Planning and implementing service management**	✔					No change, left in for context of other changes.
4.1 Plan service management (Plan)	**4.1 Plan service management (Plan)**	✔					No change, left in for context of other changes.
	4.1.1 Scope of service management		✔				New subclause heading for ISO formatting.
Management define the scope as part of their management responsibilities (and as part of the service management plan).	**Management should define the scope as part of their management responsibilities** (and as part of the service management plan).			✔			Clarification of the recommendation by change to the wording. This is not a change to the recommendation.
The scope should then be checked for suitability under **BS 15000-1.**	The scope should then be checked for suitability under **ISO/IEC 20000-1.**		✔				
NOTE Planning for operational changes is described **in 8.2.**	NOTE Planning for operational changes is described **in 9.2**		✔				Change to the cross-reference reflecting the re-numbering of clauses.
	4.1.2 Planning approaches		✔				New subclause heading for ISO formatting.
Multiple service management plans may be used in place of one large plan or programme.	Multiple service management plans may be used in place of one large plan or programme.	✔					No change, left in for context of other changes.

Table 7 – The Part 2 differences (continued)

BS 15000-2	ISO/IEC 20000-2	N/C	Form	Edit	Rec	Req	Commentary
	4.1.3 Events to be considered		✔				New subclause heading for ISO/IEC formatting.
The service management plan should cater for service management process and service changes triggered by events such as:.....	The service management plan should cater for service management process and service changes triggered by events such as:......		✔				Minor formatting change of numeric list converted to alpha list.
	4.1.4 Scope and contents of the plan		✔				New subclause heading for ISO/IEC formatting.
A service management plan should define:	A service management plan should define:		✔				Change to list numbering during ISO/IEC formatting.
i) the scope of **service management within the organization;**	a) the scope of the **service provider's service management;**			✔			Clarification from use of standard terms.
ii) the objectives **that** are to be achieved;	b) the objectives **and requirements that** are to be achieved by service management;				✔		Change in the recommended scope of the service management plan.
iii)the resources and **facilities** necessary to achieve the defined objectives;	c) the resources, **facilities and budgets** necessary to achieve the defined objectives;				✔		Change in the scope of the service management plan, now recommends including budgets.
iv)the framework of management roles and responsibilities, including the **management of third-party suppliers;**	d) the framework of management roles and responsibilities, including the **senior responsible owner, process owners** and management of **suppliers;**				✔		Change in the scope of the service management plan, now recommends including senior responsible owner and process owner roles.
v) the interfaces between service management processes and the manner in which **processes** are to be coordinated;	e) the interfaces between service management processes and the manner in which the **activities and/or processes** are to be co-ordinated;				✔		Change in the recommended level of detail to be considered for interface details.

Table 7 – The Part 2 differences (continued)

BS 15000-2	ISO/IEC 20000-2	N/C	Form	Edit	Rec	Req	Commentary
vi) the approach to be taken in identifying, assessing and **managing risks so the defined objectives are achieved;**	f) the approach to be taken in identifying, assessing and **managing issues and risks to the achievement of the defined objectives;**				✔		Small change to the recommendations from including issues as well as risks.
vii) a resource schedule expressed in terms of the dates on which funds, skills, and resources should be available;	g) a resource schedule expressed in terms of the dates on which funds, skills, and resources should be available;		✔				Minor formatting change, alpha bullets instead of numeric.
viii) the approach to changing the plan and the service defined by the plan;	h) the approach to changing the plan and the service defined by the plan;		✔				Minor formatting change, alpha bullets instead of numeric.
ix) how the **organization** will demonstrate continuing quality control (e.g. interim audits).	i) how the **service provider** will demonstrate continuing quality control (e.g. interim audits);			✔			Clarification from use of a standard term.
	j) the processes that are to be executed;				✔		New recommendation for the contents of a plan.
	k) tools as appropriate to support the processes.				✔		New recommendation for the contents of a plan.
4.2 Implement service management and provide the services (Do)	**4.2 Implement service management and provide the services (Do)**	✔					No change, left in for context of other changes.
Objective: To implement the service management ***plan and achieve the defined service objectives.***	*Objective: To implement the service management* ***objectives and plan.***				✔		Change to the objective which does not change the requirements, as the objectives are an explanation/advice.
Attainment of best practice service management processes capable of meeting the requirements of **BS 15000** will not be achieved if the original services do not meet the requirements outlined for the implementation in **BS 15000 Part 1.**	Attainment of best practice service management processes capable of meeting the requirements of **ISO/IEC 20000** will not be achieved if the original services do not meet the requirements outlined for the implementation in **ISO/IEC 20000-1**.		✔				

Table 7 – The Part 2 differences (continued)

BS 15000-2	ISO/IEC 20000-2	N/C	Form	Edit	Rec	Req	Commentary
Once implemented the service and service management processes **should be maintained with the same level of rigour used in the original service management plan.**	Once implemented the service and service management processes **should be maintained.**			✔			References to 'level of rigour' have been deleted. The 'level of rigour' added no value to this recommendation as the level of rigour for the processes is defined by customer and business needs, the requirements for improvements and for the individual service management processes.
4.3 Monitoring, measuring and reviewing (Check)	**4.3 Monitoring, measuring and reviewing (Check)**	✔					No change, left in for context of other changes.
4.3.1 Measurement and analysis			✔				Subclause heading deleted as part of aligning Parts 1 and 2.
The **organization** should plan and implement the monitoring, measurement and analysis of the service, the service management processes and associated systems.	The **service provider** should plan and implement the monitoring, measurement, analysis and review of the service, the service management processes and associated systems.			✔			Clarification form the use of a standard term.
The results of the analysis should provide input to the service improvement programme.				✔			Deleted as this recommendation was superfluous when the whole clause is part of the PDCA cycle.
Items that should be monitored, measured and **analysed** include:	Items that should be monitored, measured, and **reviewed** include:				✔		By changing 'analysed' to 'reviewed' this has changed the recommendations for the whole sequence of events in a) to f).
a) achievement against defined service targets;	**a) achievement against defined service targets;**				✔		
b) customer satisfaction;	**b) customer satisfaction;**				✔		
c) resource utilisation;	**c) resource utilisation;**				✔		

Table 7 – The Part 2 differences (continued)

BS 15000-2	ISO/IEC 20000-2	N/C	Form	Edit	Rec	Req	Commentary
d) trends;	d) trends;				✔		
e) major non-conformities;	e) major non-conformities.				✔		
f) results of reviews.					✔		
	The results of the analysis should provide input to a plan for improving the service.				✔		This is a new recommendation arising from the change to the previous recommendation.
As well as service management activities on measurement and analysis senior management may need to make use of internal audits and other checks.	As well as service management activities on measurement and analysis senior management may need to make use of internal audits and other checks.	✔					No change, left in for context of other changes.
When deciding the frequency of such internal audits and checks, the degree of risk involved in a process, its frequency of operation and its past history of problems are among the factors that should be taken into account.	When deciding the frequency of such internal audits and checks, the degree of risk involved in a process, its frequency of operation and its past history of problems are among the factors that should be taken into account.	✔					No change, left in for context of other changes.
In common with external certification and re-certification audits, internal audits and checks should be planned, carried out competently, and recorded as they would be for an independent external audit.	Internal audits and checks should be planned, carried out competently and recorded.			✔			The reference to external audits is not appropriate in an ISO/IEC standard of this type. Also the recommendation is not changed by the deletion of the text.
4.4 Continuous improvement (Act)	**4.4 Continual improvement (Act)**			✔			This is correction of the word 'continuous' by use of 'continual' in line with other management system standards. This is not a change to the intentions of this clause.

Table 7 – The Part 2 differences (continued)

BS 15000-2	ISO/IEC 20000-2	N/C	Form	Edit	Rec	Req	Commentary
4.4.1 Policy	***4.4.1 Policy***	✔					No change, left in for context of other changes.
There should be effective liaison within the **supplier's own management structure,** customers **and other suppliers** on matters affecting service quality and customer requirements.	There should be effective liaison within the **service provider's own management structure,** customers and **the service provider's suppliers** on matters affecting service quality and customer requirements.			✔			Clarification from use of a standard term.
4.4.2 Planning for service improvements	***4.4.2 Planning for service improvements***	✔					No change, left in for context of other changes.
Before implementing a **service improvement plan**, service quality and levels should be recorded as a baseline against which the actual improvements can be compared.	Before implementing **a plan for improving the service,** service quality and levels should be recorded as a baseline against which the actual improvements can be compared.			✔			Clarification from use of a standard term.
Service improvement targets should be measurable, linked to business objectives and documented **in the plan**.	Service improvement targets should be measurable, linked to business objectives and documented **in a plan**.			✔			Clarification from use of a standard term.
4.4.3 Planning and implementing new or changed services	**5 Planning and implementing new or changed services**			✔			Clause heading number change, aligning Parts 1 and 2.
	5.1 Topics for consideration		✔				New clause heading for ISO formatting.
Planning for new or changed services should include reviewing:	Planning for new or changed services should include reviewing:	✔					No change, left in for context of other changes.
a) budgets;	a) budgets;	✔					
b) staff resources;	b) staff resources;	✔					
c) existing service levels,	c) existing service levels,	✔					

Table 7 – The Part 2 differences (continued)

BS 15000-2	ISO/IEC 20000-2	N/C	Form	Edit	Rec	Req	Commentary
d) SLAs and other targets or service commitments;	d) SLAs and other targets or service commitments;	✔					No change, left in for context of other changes.
e) existing service management processes, procedures and documentation;	e) existing service management processes, procedures and documentation;	✔					
f) the scope of service management, including the implementation of service management processes previously excluded from the scope.	f) the scope of service management, including the implementation of service management processes previously excluded from the scope.	✔					
	5.2 Change records		✔				New clause heading for ISO formatting.
All service changes should be reflected in Change Management records.	All service changes should be reflected in Change Management records.	✔					
This includes plans for:	This includes plans for:		✔				Change to alpha list from original numeric list, as part of ISO/IEC formatting.
1) staff recruitment/retraining;	a) staff recruitment/retraining;		✔				
2) relocation;	b) relocation;		✔				
2) relocation;	b) relocation;		✔				
3) user training;	c) user training;		✔				
4) communications about the changes;	d) communications about the changes;		✔				
5) changes to the nature of the technology supported;	e) changes to the nature of the technology supported;		✔				
6) formal closure of services.	f) formal closure of services.		✔				

Table 7 – The Part 2 differences (continued)

BS 15000-2	ISO/IEC 20000-2	N/C	Form	Edit	Rec	Req	Commentary
5 Service delivery processes	**6 Service delivery processes**		✔				Clause heading number change, aligning Parts 1 and 2.
5.1 Service level management	**6.1 Service level management**		✔				Clause heading number change, aligning Parts 1 and 2.
5.1.1 Service catalogue	***6.1.1 Service catalogue***		✔				Subclause heading number change.
5.1.2 Service level agreements (SLAs)	***6.1.2 Service level agreements (SLAs)***		✔				Subclause heading number change aligning Parts 1 and 2.
The minimum content that should be in an SLA or that can be directly referenced from an SLA is:	The minimum content that should be in an SLA or that can be directly referenced from an SLA is:	✔					No change, left in for context of other changes.
a) brief service description;	a) brief service description;	✔					
b) validity period and/or SLA change control mechanism;	b) validity period and/or SLA change control mechanism;	✔					
c) authorization details;	c) authorization details;	✔					
d) brief description of communications, including reporting;	d) brief description of communications, including reporting;	✔					
e) contact details of people authorized to act in emergencies, to participate in incidents and problem correction, recovery or workaround;	e) contact details of people authorized to act in emergencies, to participate in incidents and problem correction, recovery or workaround;	✔					
f) service hours, e.g. 09:00 h to 17:00 h, date exceptions (e.g. weekends, public holidays), critical business periods and out of hours cover;	f) service hours, e.g. 09:00 h to 17:00 h, date exceptions (e.g. weekends, public holidays), critical business periods and out of hours cover;	✔					

Table 7 – The Part 2 differences (continued)

BS 15000-2	ISO/IEC 20000-2	N/C	Form	Edit	Rec	Req	Commentary
g) scheduled and agreed interruptions, including notice to be given, number per period;	g) scheduled and agreed interruptions, including notice to be given, number per period;	✔					
h) customer responsibilities, e.g. security;	h) customer responsibilities, e.g. security;	✔					
i) service provider liability and obligations e.g. security;	i) service provider liability and obligations e.g. security;	✔					
j) impact and priority guidelines;	j) impact and priority guidelines;	✔					
k) escalation and notification process;	k) escalation and notification process;	✔					
l) complaints procedure;	l) complaints procedure;	✔					This is a recommendation that supports the complaints process that is now required by ISO/IEC 20000-1. As this Part 2 clause supports Part 1 it has been left as a complaints procedure.
m) service targets;	m) service targets;	✔					No change, left in for context of other changes.
n) workload limits (upper and lower), e.g. the ability of the service to support the agreed number of **customers/volume of work**, system throughput;	n) workload limits (upper and lower), e.g. the ability of the service to support the agreed number of **users/volume of work**, system throughput;			✔			Clarification from corrected reference to the change management process. This is not a change to the recommendation.
o) high level financial management details, e.g. charge codes etc;	o) high level financial management details, e.g. charge codes etc;	✔					No change, left in for context of other changes.
p) action to be taken in the event of a service interruption;	p) action to be taken in the event of a service interruption;	✔					

Table 7 – The Part 2 differences (continued)

BS 15000-2	ISO/IEC 20000-2	N/C	Form	Edit	Rec	Req	Commentary
q) housekeeping procedures;	q) housekeeping procedures;	✔					No change, left in for context of other changes.
r) glossary of terms;	r) glossary of terms;	✔					
s) supporting and related services;	s) supporting and related services;	✔					
t) any exceptions to the terms given in the SLA.	t) any exceptions to the terms given in the SLA.	✔					
NOTE 2 Volatile information, or information common to many SLAs (such as contact details) can be referenced from the SLA without impacting the quality of SLM processes as long as the referenced documents are also **<u>under change contro</u>**l.	NOTE 2 Volatile information, or information common to many SLAs (such as contact details) can be referenced from the SLA without impacting the quality of SLM processes as long as the referenced documents are also **<u>under the control of the change management process.</u>**			✔			Clarification from corrected reference to the change management process in line with the intent of the original sentence in BS 15000. This is not a change to the recommendation.
NOTE 3 Continuity plan and details of **<u>financial management</u>** are normally referenced from the SLA.	NOTE 3 Continuity plan and details of **<u>accounting and budgeting</u>** are normally referenced from the SLA.			✔			The revised wording reflects the scope of clause 6.4 and is not a change to the recommendations. A service provider may however choose to add links to financial processes such as changing in their SLAs.
<u>5.1.3</u> Service level management (SLM) process	***<u>6.1.3</u> Service level management (SLM) process***		✔				Subclause heading number change.
The SLM process should manage and coordinate contributors of the service levels, to include:	The SLM process should manage and coordinate contributors of the service levels, to include:	✔					No change, left in for context of other changes.
a) agreement of the service requirements and expected service workload characteristics;	a) agreement of the service requirements and expected service workload characteristics;	✔					No change, left in for context of other changes.
b) agreement of service targets;	b) agreement of service targets;	✔					No change, left in for context of other changes.

Table 7 – The Part 2 differences (continued)

BS 15000-2	ISO/IEC 20000-2	N/C	Form	Edit	Rec	Req	Commentary
c) measurement and reporting of the service levels achieved, workloads and an explanation if the agreed targets are not met **(see 5.3);**	c) measurement and reporting of the service levels achieved, workloads and an explanation if the agreed targets are not met **(see 6.2);**		✔				Change to cross-reference due to re-numbering of clauses. This does not change recommendations.
d) **instigation** of corrective action;	d) **initiation** of corrective action;			✔			Change from 'instigation' to 'initiation' to assist understanding by an international audience. This is not a change to the original intent of the recommendation.
e) **input to the service improvement programme**.	e) input to **a plan for improving the service.**			✔			
The SLM process should work closely with business relationship and supplier **management.**	The SLM process should work closely with the business relationship and supplier **management processes.**			✔			Clarification of the link being to a process, not to a team of people, in line with the process basis of both parts of the standard.
5.1.4 ***Supporting service agreements***	***6.1.4*** ***Supporting service agreements***		✔				Subclause heading number change.
The supporting services on which the delivered service depends should be documented and agreed with each **service supplier.**	The supporting services on which the delivered service depends should be documented and agreed with each **supplier.**			✔			Clarification of the recommendation, not a change to the recommendation.
	This includes internal groups providing part of the service provider's service.			✔			
These are typically called operational level agreements (OLAs) or underpinning contracts.				✔			

Table 7 – The Part 2 differences (continued)

BS 15000-2	ISO/IEC 20000-2	N/C	Form	Edit	Rec	Req	Commentary
<u>5.2</u> Service continuity and availability management	**<u>6.3</u> Service continuity and availability management**		✔				**DIFFERENT SEQUENCE – THIS IS NOW THE 3rd CLAUSE IN CLAUSE 6, IN BS 15000-2 IT WAS THE 2nd CLAUSE. THIS ALIGNS PARTS 1 AND 2.**
*Objective: To ensure that agreed obligations to customers can be met **<u>in all circumstances from normal through to a major loss of service.</u>***	*Objective: To ensure that agreed service continuity and availability commitments to customers can be met **<u>in all circumstances.</u>***			✔			Change in wording aligning Parts 1 and 2.
<u>5.2.1</u> General	***<u>6.3.1</u> General***		✔				Subclause heading number change.
Requirements should include access rights and response times as well as **<u>end to end</u>** availability of system components.	Requirements should include access rights and response times as well as **<u>end-to-end</u>** availability of system components.		✔				Trivial change from use of hyphens.
<u>5.2.2</u> Availability monitoring and activities	***<u>6.3.2</u> Availability monitoring and activities***		✔				Subclause heading number change.
Availability management should:	Availability management should:	✔					No change, left in for context of other changes.
a) monitor and record availability of the service	a) monitor and record availability of the service	✔					
b) maintain accurate historical data;	b) maintain accurate historical data;	✔					
c) make comparisons with requirements defined in SLAs to identify non-conformance to the agreed availability targets;	c) make comparisons with requirements defined in SLAs to identify non-conformance to the agreed availability targets;	✔					
d) document and review non-conformance;	d) document and review non-conformance;	✔					
e) predict future availability;	e) predict future availability;	✔					

Table 7 – The Part 2 differences (continued)

BS 15000-2	ISO/IEC 20000-2	N/C	Form	Edit	Rec	Req	Commentary
	<u>f)</u> **<u>where possible, potential issues should be predicted and preventive action taken.</u>**				✔		This was a requirement converted to a NOTE in the specification. The use of '*Where possible…*' is not suitable for a requirement but is suitable as a recommendation in Part 2.
<u>5.2.3</u> Service continuity strategy	***<u>6.3.3</u> Service continuity strategy***		✔				Subclause heading number change.
5.2.4 Service continuity planning and testing	***6.3.4 Service continuity planning and testing***		✔				Subclause heading number change.
Service continuity plans and related documents (e.g. contracts) should be linked to: **<u>1)</u> <u>the change management process.</u>** **<u>2)</u> <u>the contract management process.</u>**	Service continuity plans and related documents (e.g. contracts) should be linked to **<u>the change management process and the contract management process.</u>**		✔				Change for ISO formatting. This does not change the recommendations.
Testing should be undertaken at a frequency and rigour sufficient to gain assurance that continuity plans are effective, and remain so in the face of changing systems, processes, personnel and business needs.	Testing should be undertaken at a frequency sufficient to gain assurance that continuity plans are effective, and remain so in the face of changing systems, processes, personnel and business needs.	✔					No change, left in for context of other changes.
Testing should be a joint involvement between customer and service provider based upon an agreed set of objectives.	Testing should be a joint involvement between customer and service provider based upon an agreed set of objectives.	✔					No change, left in for context of other changes.
Test failures should be documented and reviewed to **<u>input to the service improvement plan.</u>**	Test failures should be documented and reviewed to **<u>input to a plan for improving the service.</u>**			✔			Clarification from standardization of terms.

Table 7 – The Part 2 differences (continued)

BS 15000-2	ISO/IEC 20000-2	N/C	Form	Edit	Rec	Req	Commentary
<u>5.3</u> **Service reporting**	<u>6.2</u> **Service reporting**		✔				**SEQUENCING CHANGE – THIS CLAUSE IS NOW THE 2nd SUBCLAUSE IN CLAUSE 6, COMING BEFORE THE SUBCLAUSE ON SERVICE CONTINUITY AND AVAILABILITY MANAGEMENT. THIS ALIGNS PARTS 1 AND 2.**
<u>5.3.1 Policy</u>	NOTE The success of all service management processes is dependent on the use of the information provided in service reports.		✔				Subclause heading moved from before the NOTE to after the NOTE.
NOTE The success of all service management processes is dependent on the use of the information provided in service reports.	***<u>6.2.1 Policy</u>***		✔				
Where there are multiple service providers, **<u>suppliers and third-party suppliers</u>** the reports should reflect the relationships between suppliers.	Where there are **<u>multiple suppliers, lead suppliers and sub-contracted suppliers</u>** the reports should reflect the relationships between suppliers.			✔			Clarification from standardization of terms supplier, lead supplier and sub-contracted supplier. This is not a change to the original intent of this clause, but this clause may have been misunderstood in BS 15000-2.
For example, a lead supplier should report on the whole of the service they provide, including any services by **<u>third-party suppliers</u>** that they manage as part of the customer's service.	For example, a lead supplier should report on the whole of the service they provide, including any services by **<u>sub-contracted suppliers</u>** that they manage as part of the customer's service.			✔			Clarification from standardization of terms. This is not a change to the intent of this clause.
<u>5.3.2</u> Purpose and quality checks on service reports	***<u>6.2.2</u> Purpose and quality checks on service reports***		✔				Subclause heading number change.
<u>5.3.3</u> Service reports	***<u>6.2.3</u> Service reports***		✔				Subclause heading number change.

Table 7 – The Part 2 differences (continued)

BS 15000-2	ISO/IEC 20000-2	N/C	Form	Edit	Rec	Req	Commentary
The **lead supplier** should produce reports for customers and management covering:…………..	The **service provider** should produce reports for customers and management covering:………..			✔			Clarification from standardization of a term.
5.4 Budgeting and accounting for IT services	**6.4 Budgeting and accounting for IT services**		✔				Clause heading number change, aligning Parts 1 and 2.
	6.4.1 General		✔				New subclause heading for ISO formatting.
This section covers budgeting and accounting for IT services.	This section covers budgeting and accounting for IT services.	✔					No change, left in for context of other changes.
In practice, many **organizations** will be involved in charging for such services.	In practice, many **service providers** will be involved in charging for such services.			✔			Clarification from standardization of a term.
However, since charging is an optional activity, it is not covered by the standard.	However, since charging is an optional activity, it is not covered by the standard.	✔					No change, left in for context of other changes.
Organizations are recommended that where charging is in use, the mechanism for doing so is fully defined and understood by all parties.	**Service providers** are recommended that where charging is in use, the mechanism for doing so is fully defined and understood by all parties.			✔			Clarification from standardization of a term.
Responsibility for many of the financial decisions will lie outside the sphere of the service management arena and the requirements for what financial information is to be provided, in what form and at what frequencies may be dictated from outside.	Responsibility for many of the financial decisions will lie outside the sphere of the service management arena and the requirements for what financial information is to be provided, in what form and at what frequencies may be dictated from outside.	✔					No change, left in for context of other changes.
The provisions of this section are focused on the practices that should be followed to satisfy the requirements of the standard.	The provisions of this section are focused on the practices that should be followed to satisfy the requirements of the standard.	✔					No change, left in for context of other changes.

Table 7 – The Part 2 differences (continued)

BS 15000-2	ISO/IEC 20000-2	N/C	Form	Edit	Rec	Req	Commentary
However, wider requirements should also be taken into account as they will impact on some of the policies and procedures defined.	However, wider requirements should also be taken into account as they will impact on some of the policies and procedures defined.	✔					No change, left in for context of other changes.
All accounting practices used should be aligned to the wider accountancy practices of the **organization.**	All accounting practices used should be aligned to the wider accountancy practices of the whole of the **service provider's organization.**			✔			Clarification from standardization of a term.
***5.4.1* Policy**	***6.4.2* Policy**		✔				Subclause heading number change.
The policy should also define the detail to which budgeting and accounting are performed, taking into consideration the:	The policy should also define the detail to which budgeting and accounting are performed, taking into consideration the:	✔					
a) cost types to be accounted for;	a) cost types to be accounted for;	✔					No change, left in for context of other changes.
b) apportionment of overhead costs, e.g. flat rate, fixed percentage, or based on the size of the variable element;	b) apportionment of overhead costs, e.g. flat rate, fixed percentage, or based on the size of the variable element;	✔					
c) granularity of the **customer's business against which charges are levied,** e.g. business unit as one unit, subdivided into department, or by locations;	c) granularity of the **customer's business,** e.g. business unit as one unit, subdivided into department, or by locations;			✔			Clarification of the meaning from re-wording.
d) rules governing the handling of variances against budgets, e.g. size of variance that will be escalated to senior management; **links to service level management.**	d) rules governing the handling of variances against budgets, e.g. size of variance that will be escalated to senior management;			✔			The original bullet d) in BS 15000-2 has been split into two bullets (as originally intended). This results in a minor change to the recommendations. This clause may have been misunderstood in BS 15000-2.

Table 7 – The Part 2 differences (continued)

BS 15000-2	ISO/IEC 20000-2	N/C	Form	Edit	Rec	Req	Commentary
	e) **links to service level management.**			✔			Bullet e) split from bullet d), see previous comments.
The level of investment in budgeting and accounting processes should be based on the needs of both the **customer and supplier** for financial detail as defined in the policy.	The level of investment in budgeting and accounting processes should be based on the needs of the **customers, service provider and suppliers** for financial detail as defined in the policy.				✔		Clarification of the recommendation and extension to all three parties in a supply chain.
NOTE Service providers operating in a commercial environment might need to invest considerably more time and effort in their financial management. Conversely, for **organizations** where simple identification of costs is sufficient the financial management can be much simpler.	NOTE Service providers operating in a commercial environment might need to invest considerably more time and effort in their financial management. Conversely, for **service providers** where simple identification of costs is sufficient the financial management can be much simpler.			✔			Clarification from standardization of a term.
Budgeting and accounting should be performed by all **organizations,** whatever their other policies on financial management.	Budgeting and accounting should be performed by all **service providers,** whatever their other policies on financial management.			✔			Clarification from standardization of a term.
***5.4.2* Budgeting**	***6.4.3* Budgeting**		✔				Subclause heading number change.
***5.4.3* Accounting**	***6.4.4* Accounting**		✔				Subclause heading number change.
5.5 Capacity management	**6.5 Capacity management**		✔				Clause heading number change, aligning Parts 1 and 2.
*Objective: To ensure that the **organization** has, at all times, sufficient capacity to meet the current and future agreed **demands of the business.***	*Objective: To ensure that the **service provider** has, at all times, sufficient capacity to meet the current and future agreed **demands of customer's business needs.***			✔			Clarification of the objectives from re-wording. Part 1 and Part 2 objectives have also been aligned.

Table 7 – The Part 2 differences (continued)

BS 15000-2	ISO/IEC 20000-2	N/C	Form	Edit	Rec	Req	Commentary
5.6 Information security management	**6.6 Information security management**		✔				Clause heading number change, aligning Parts 1 and 2.
	6.6.1 General		✔				New subclause heading for ISO formatting.
NOTE: Information security is the result of a system of policies and procedures designed to identify, control and protect information and any equipment used in connection with its storage, transmission and processing.	**Information security** is the result of a system of policies and procedures designed to identify, control and protect information and any equipment used in connection with its storage, transmission and processing.		✔				A NOTE has been converted to normal text. The status of a NOTE in part 2 means that conversion to normal text has not changed the recommendations, as this is a code of practice.
	The service provider's staff with **specialis**t information security roles should be conversant with **ISO/IEC 17799*, Information technology - Security techniques - Code of practice for information security management***				✔		Moved from subclause 5.6.1 in BS 15000-2 column. Re-worded for use in an international standard. *'Information security'* roles has been re-worded as *'specialist information security roles'*. This is a minor change to the recommendations.
BS 7799-1 provides guidance on information security management.				✔			This part of the NOTE has been deleted as referring to a national standard is not appropriate for an international standard. ISO/IEC 17799 is referred to in the second sentence of 6.6.1, ISO/IEC 20000-2, above.

Table 7 – The Part 2 differences (continued)

BS 15000-2	ISO/IEC 20000-2	N/C	Form	Edit	Rec	Req	Commentary
Implementation of the requirements in this specification may not satisfy all the requirements that are necessary to obtain certification against BS 7799-2.				✔			This part of the NOTE has been deleted as reference to a national standard is not appropriate for an international standard. This part of the NOTE in BS 15000-2 was advice, not a recommendation.
Organizations certified to BS 7799 will satisfy the security requirements within service management.					✔		This part of the NOTE has been deleted as reference to a national standard is not appropriate for an international standard. Deletion changes the recommendations but does not change the requirements. Service providers certified to BS 7799 were still required to be audited against the requirements of clause 5.6 of BS 15000-1 and now to be audited against the requirements of clause 6.6 in ISO/IEC 20000-1.
***5.6.1** Identifying and classifying information assets*	***6.6.2** Identifying and classifying information assets*		✔				Subclause heading number change.
The service provider's staff with information security roles should be conversant with **BS 7799, the British Standard for Information Security Management.**					✔		Moved to subclause level, see 6.6.1 in ISO/IEC 20000-2. *'Information security'* roles has been re-worded as *'specialist information security roles'*.

Table 7 – The Part 2 differences (continued)

BS 15000-2	ISO/IEC 20000-2	N/C	Form	Edit	Rec	Req	Commentary
***5.6.2** Security risk assessment*	***6.6.3** Security risk assessment practices*		✔				Subclause heading number change and small heading change.
Security risk assessment should: a) be performed at agreed intervals; b) be recorded; c) be maintained during changes (of changing business needs, processes and configurations); d) help understanding of what could impact a managed service; e) inform decisions regarding the types of controls to be operated.	Security risk assessment should: a) be performed at agreed intervals; b) be recorded; c) be maintained during changes (of changing business needs, processes and configurations); d) help understanding of what could impact a managed service; e) inform decisions regarding the types of controls to be operated.	✔					No change, left in to clarify where the new heading has been inserted.
	6.6.4 Risks to information assets		✔				New subclause heading inserted for ISO formatting.
Risks to information assets should be assessed by reference to:	Risks to information assets should be assessed by reference to:	✔					Minor formatting change, numeric bullets have become alpha bullets.
1) their nature (e.g. software malfunction, operating errors, communications failure);	a) their nature (e.g. software malfunction, operating errors, communications failure);		✔				
2) likelihood;	b) likelihood;		✔				
3) potential business impact;	c) potential business impact;		✔				
4) past experience.	d) past experience.		✔				

Table 7 – The Part 2 differences (continued)

BS 15000-2	ISO/IEC 20000-2	N/C	Form	Edit	Rec	Req	Commentary
	6.6.5 Security and availability of information		✔				New subclause heading inserted for ISO formatting.
In assessing risks, due regard should be paid to the following:	In assessing risks, due regard should be paid to the following:	✔					No change, left in for context of other changes.
i) disclosure of sensitive information to unauthorized parties;	a) disclosure of sensitive information to unauthorized parties;		✔				Minor formatting change of numeric to alpha bullets.
ii) inaccurate, incomplete or invalid (e.g. fraudulent) information;	b) inaccurate, incomplete or invalid (e.g. fraudulent) information;		✔				
iii)information being unavailable for use (e.g. due to power failure);	c) information being unavailable for use (e.g. due to power failure);		✔				
iv)physical damage to, or destruction of equipment necessary to provide services.	d) physical damage to, or destruction of equipment necessary to provide services.		✔				
Account should also be taken of information security policy objectives, the need to meet customers' specified security requirements (e.g. availability levels), **and statutory or regulatory requirements that apply (e.g. the Data Protection Act 1998 [1]).**	Account should also be taken of information security policy objectives, the need to meet customers' specified security requirements (e.g. availability levels), **and statutory or regulatory requirements that apply.**				✔		Deletion of the reference to UK legislation not appropriate for an international standard. This is a small change in the recommendation but the wording still reflects the original intent.
NOTE See also PD 3002, Guide to Risk Assessment and Risk Management.					✔		Note has been deleted as it refers to a national publication. This is a small change to the recommendations.
5.6.3 Controls	***6.6.6 Controls***		✔				Subclause heading number change, aligning Parts 1 and 2.

Table 7 – The Part 2 differences (continued)

BS 15000-2	ISO/IEC 20000-2	N/C	Form	Edit	Rec	Req	Commentary
In addition to other controls that might be justified and advice contained elsewhere in **<u>BS</u> <u>15000</u>** (e.g. on service continuity), service providers should operate the following controls as a matter of good information security management practice:......	In addition to other controls that might be justified and advice contained elsewhere in **<u>this part of ISO/IEC</u> <u>20000</u>** (e.g. on service continuity), service providers should operate the following controls as a matter of good information security management practice:.......		✔				Change to how the standard is referred to in line with ISO/IEC practices.
	6.6.7 Documents and records		✔				Subclause heading inserted for ISO formatting.
Records should be analysed periodically to provide management with information on:	Records should be analysed periodically to provide management with information on:	✔					No change, left in for context of other changes.
1) effectiveness of information security policy;	a) effectiveness of information security policy;		✔				Minor formatting change of numeric to alpha bullets.
2) emerging trends in information security incidents;	b) emerging trends in information security incidents;		✔				Minor formatting change of numeric to alpha bullets.
3) **<u>input to service improvement plans;</u>**	c) **<u>input to a plan for improving the service</u>**;			✔			Clarification from standardization of terms.
4) control over access to information, assets, a nd systems.	d) control over access to information, assets, and systems.		✔				Minor formatting change of numeric to alpha bullets.
The information security management system should be reliably documented.				✔			Recommendation deleted as it added no value to a subclause that was on the subject of recommended documentation.
<u>6</u> Relationship processes	**<u>7</u> Relationship processes**		✔				Clause heading number change, aligning Parts 1 and 2.
<u>6.1</u> General	**<u>7.1</u> General**		✔				Clause heading number change, aligning Parts 1 and 2.

Table 7 – The Part 2 differences (continued)

BS 15000-2	ISO/IEC 20000-2	N/C	Form	Edit	Rec	Req	Commentary
Relationship processes describe the two related aspects of Supplier Management and **Business** Management.	Relationship processes describe the two related aspects of Supplier Management and **Business Relationship** Management.			✔			Clarification from standardization of terms.
This standard addresses the role of a service provider, who logically fills a role between suppliers, delivering goods or services to the service provider, and the customer who receives services.	This standard addresses the role of a service provider, who logically fills a role between suppliers, delivering goods or services to the service provider, and the customer who receives services.	✔					No change, left in for context of other changes.
Both the suppliers and customer may be internal or external to the service provider's organization.	Both the suppliers and customer may be internal or external to the service provider's organization.	✔					No change, left in for context of other changes.
External relationships will be formalized via a contract,....	External relationships will be formalized via a contract.		✔				
..... internal ones by a service or operational level agreement.	**Internal relationships will be formalized by a service or internal under-pinning agreement often referred to as an operational level agreement.**			✔			Clarification of the recommendation from re-wording. This is not a change to requirements or to recommendations.
Figure 2 shows a simplified representation of the relationships.	Figure 2 shows a simplified representation of the relationships.	✔					

Table 7 – The Part 2 differences (continued)

BS 15000-2	ISO/IEC 20000-2	N/C	Form	Edit	Rec	Req	Commentary
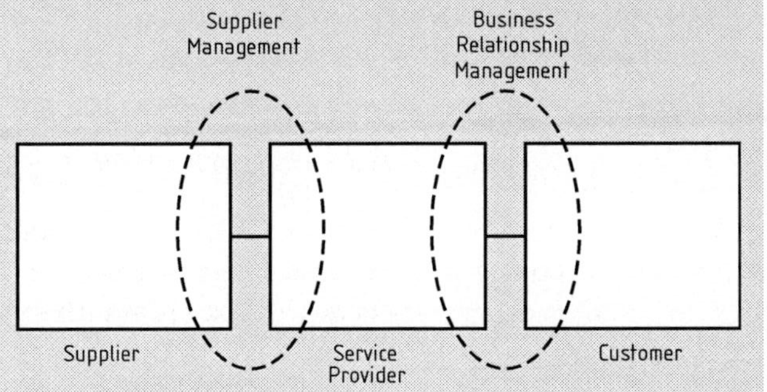	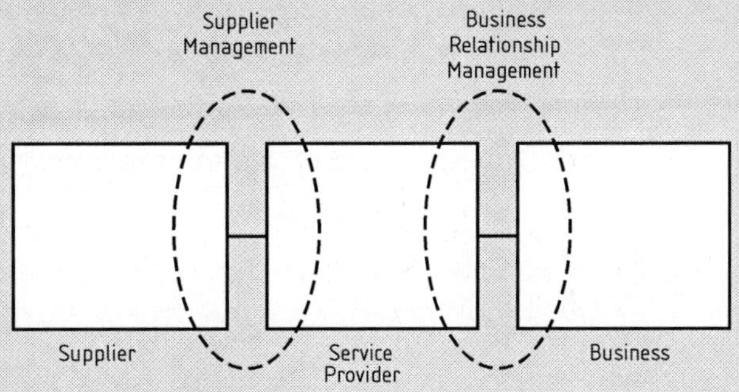			✔			Clarification from standardization of terms (in line with the Figure in P art 1), with Customer changed to Business. This is not a change to the intent of this clause and does not change the recommendation.
Figure 2 — Relationship processes	**Figure 2 — Relationship processes**	✔					
6.2 Business relationship management	**7.2 Business relationship management**		✔				Clause heading number change, aligning Parts 1 and 2.
6.2.1 Service reviews	***7.2.1 Service reviews***		✔				Subclause heading number change.
6.2.2 Service complaints	***7.2.2 Service complaints***		✔				Subclause heading number change.
The service provider and customer(s) should agree on a formal complaints procedure so that there is no ambiguity on what constitutes a complaint and how it should be handled.	The service provider and customer(s) should agree on a formal complaints procedure so that there is no ambiguity on what constitutes a complaint and how it should be handled.	✔					This is a recommendation that supports the complaints process that is now required by ISO/IEC 20000-1. As this Part 2 clause supports Part 1 it has been left as a complaints procedure.

Table 7 – The Part 2 differences (continued)

BS 15000-2	ISO/IEC 20000-2	N/C	Form	Edit	Rec	Req	Commentary
Service providers should periodically analyse the record of complaints to identify trends and report this analysis to customers.	Service providers should periodically analyse the record of complaints to identify trends and report this analysis to customers.	✔					No change, left in for context of other changes.
The results of such analysis should be used where appropriate to **inform the service improvement plan**.	The results of such analysis should be used where appropriate to **inform a plan for improving the service.**			✔			Clarification from standardization of terms.
***6.2.3* Customer satisfaction measurement**	***7.2.3* Customer satisfaction measurement**		✔				Subclause heading number change.
The results and conclusions of customer satisfaction surveys should be discussed with the customer.	The results and conclusions of customer satisfaction surveys should be discussed with the customer.	✔					No change, left in for context of other changes.
An action plan should be agreed, **input to the service improvement plan** and progress reported back to the customer.	An action plan should be agreed, **input to a plan for improving the service** and progress reported back to the customer.			✔			Clarification from standardization of a term.
Compliments about the service should be documented and reported to the service delivery team.	Compliments about the service should be documented and reported to the service delivery team.	✔					No change, left in for context of other changes.
6.3 Supplier management	**7.3 Supplier management**		✔				Clause heading number change, aligning Parts 1 and 2.
*Objective: To manage the **service provider(s)** to ensure the provision of seamless, quality services.*	*Objective: To manage **suppliers** to ensure the provision of seamless, quality services.*			✔			Clarification from standardization of terms.
***6.3.1* Introduction**	***7.3.1* Introduction**		✔				Subclause heading number change.
***6.3.2* Contract management**	***7.3.2* Contract management**		✔				Subclause heading number change.

Table 7 – The Part 2 differences (continued)

BS 15000-2	ISO/IEC 20000-2	N/C	Form	Edit	Rec	Req	Commentary
The **customer** should appoint a manager responsible for contracts and agreements with suppliers.	The **service provider** should appoint a manager responsible for contracts and agreements with suppliers.			✔			Clarification from the use of a standard term.
Where a number of staff are engaged in this task, there should be a common process to ensure that information on supplier performance is observed and acted on.	Where a number of staff are engaged in this task, there should be a common process to ensure that information on supplier performance is observed and acted on.	✔					No change, left in for context of other changes.
A list of contact points within the respective **organizations** should be maintained.	A list of contact points within the respective **organizations (suppliers' and service provider's)** should be maintained.			✔			Clarification of recommendation by standardization of terms.
If a contract includes penalties or bonuses, their basis should be clearly stated and compliance to the requirements reported upon.	If a contract includes penalties or bonuses, their basis should be clearly stated and compliance to the requirements reported upon.	✔					No change, left in for context of other changes.
***6.3.3* Service definition**	***7.3.3* Service definition**		✔				Subclause heading number change.
***6.3.4* Managing multiple suppliers**	***7.3.4* Managing multiple suppliers**		✔				Subclause heading number change.
It should be clear whether the service provider is dealing with all suppliers directly or a lead supplier taking responsibility for **subcontracted** suppliers.	It should be clear whether the service provider is dealing with all suppliers directly or a lead supplier taking responsibility for **sub-contracted** suppliers.		✔				Change due to inclusion of hyphen.
The lead supplier should record the names, responsibilities and relationships between all **other suppliers,** and make this available to the service provider if required.	The lead supplier should record the names, responsibilities and relationships between all **sub-contracted** suppliers, and make this available to the service provider if required.				✔		Clarification and minor change to the recommendation in this clause.

Table 7 – The Part 2 differences (continued)

BS 15000-2	ISO/IEC 20000-2	N/C	Form	Edit	Rec	Req	Commentary
The service provider should obtain evidence that lead suppliers are formally managing subcontracted providers, guided, where appropriate, by the requirements in **BS 15000-1**.	The service provider should obtain evidence that lead suppliers are formally managing sub-contracted providers, guided, where appropriate, by the requirements in **ISO/IEC 20000-1.**		✔				
***6.3.5* Contractual disputes management**	***7.3.5* Contractual disputes management**		✔				Subclause heading number change.
***6.3.6* Contract end**	***7.3.6* Contract end**		✔				Subclause heading number change.
The contract management process **shall** include provision for the end of contract – either expected end or early end.	The contract management process **should** include provision for the end of contract – either expected end or early end.			✔			The original use of the term 'shall' was not valid in Part 2, which should have only including recommendations, not the requirements which are worded 'shall'. This a correction of the wording but not a change to the intent of the clause.
It should also provide for the transfer of the service to another **party.**	It should also provide for the transfer of the service to another **organization.**			✔			Clarification from standardization of a term.
7 Resolution processes	**8 Resolution processes**		✔				Clause heading number change, aligning Parts 1 and 2.
7.1 Background	**8.1 Background**		✔				Clause heading number change, aligning Parts 1 and 2.
NOTE Incident and problem management are separate processes although they are closely linked.	Incident and problem management are separate processes although they are closely linked.			✔			Converted from a NOTE, no change to meaning.
Incident deals with the restoration of service to the users, whereas problem is concerned with identifying and removing the causes of incidents.	Incident deals with the restoration of service to the users, whereas problem is concerned with identifying and removing the causes of incidents.	✔					No change, left in for context of other changes.

Table 7 – The Part 2 differences (continued)

BS 15000-2	ISO/IEC 20000-2	N/C	Form	Edit	Rec	Req	Commentary
7.1.1 Setting priorities	*8.1.1 Setting priorities*		✔				Subclause heading number change.
7.1.2 Workarounds	*8.1.2 Workarounds*		✔				Subclause heading number change.
7.2 Incident management	8.2 Incident management		✔				Clause heading number change, aligning Parts 1 and 2.
Objective: To restore normal service as soon as possible in order to minimize business disruption.	*Objective: To restore agreed service to the business as soon as possible or to respond to service requests*			✔			Objectives of Parts 1 and 2 aligned. This is clarification and not a change to the intent of the clause.
	8.2.1 General		✔				Subclause heading number change.
NOTE The incident management process may be delivered by a service desk, which acts as the day-to-day contact with the users.	NOTE 1 The incident management process may be delivered by a service desk, which acts as the day-to-day contact with the users.		✔				Minor change to become NOTE 1, instead of being the only NOTE.
Incident management should be:	NOTE 2 Incident management should be:		✔				Minor change converting text to a NOTE. As this is a code of practice the meaning of the recommendation is unchanged irrespective of whether it is a NOTE or normal text. This assists with the alignment of Parts 1 and 2 and ISO/IEC formatting by allowing unique cross-references.
a) both a proactive and reactive process, responding to incidents that affect, or potentially could affect the service;	a) both a proactive and reactive process, responding to incidents that affect, or potentially could affect the service;		✔				
b) concerned with the restoration of the customers' service, not with determining the cause of incidents.	b) concerned with the restoration of the customers' service, not with determining the cause of incidents.		✔				

Table 7 – The Part 2 differences (continued)

BS 15000-2	ISO/IEC 20000-2	N/C	Form	Edit	Rec	Req	Commentary
The incident management process should include the following:	The incident management process should include the following:	✔					No change, left in for context of other changes.
1) call reception, recording, priority assignment, classification;	a) call reception, recording, priority assignment, classification;		✔				Minor formatting change.
2) first line resolution or referral;	b) first line resolution or referral;		✔				
3) consideration of security issues;	c) consideration of security issues;		✔				
4) incident tracking and lifecycle management;	d) incident tracking and lifecycle management;		✔				
5) incident verification and closure;	e) incident verification and closure;		✔				
6) first line customer liaison;	f) first line customer liaison;		✔				
7) escalation.	g) escalation.		✔				
7.2.1 Major incidents	*8.2.2 Major incidents*		✔				Subclause heading number change.
The process for a major incident should include a review which will inform **the service improvement programme.**	The process for a major incident should include a review which will inform **a plan for improving the service.**			✔			Clarification from standardization of terms.
7.3 Problem management	**8.3 Problem management**		✔				Clause heading number change, aligning Parts 1 and 2.
Objective: To identify and manage the underlying causes of service incidents whilst minimizing or preventing disruption to the customers.	***Objective: To minimize disruption to the business by proactive identification and analysis of the causes of incidents and by managing problems to closure.***			✔			Standardization of the objectives for Parts 1 and 2. This is not a change to requirements or recommendations. *'Service'* has now been deleted from *'service incidents'*.

Table 7 – The Part 2 differences (continued)

BS 15000-2	ISO/IEC 20000-2	N/C	Form	Edit	Rec	Req	Commentary
***7.3.1* Scope of problem management**	***8.3.1* Scope of problem management**		✔				Subclause heading number change.
***7.3.2* Initiation of problem management**	***8.3.2* Initiation of problem management**		✔				Subclause heading number change.
***7.3.3* Known errors**	***8.3.3* Known errors**		✔				Subclause heading number change.
NOTE The customer or **supplier** may decide that the resolution is too expensive or not of benefit to the business. If this is the case it should be clearly documented. The known error record should remain open however, since consequential incidents are likely to still occur and may require workarounds and/or require reassessment of the decision to resolve.	NOTE The customer or **service provider** may decide that the resolution is too expensive or not of benefit to the business. If this is the case it should be clearly documented. The known error record should remain open however, since consequential incidents are likely to still occur and may require workarounds and/or require reassessment of the decision to resolve.			✔			Clarification from standardization of a term.
***7.3.4* Problem resolution**	***8.3.4* Problem resolution**		✔				Subclause heading number change.
***7.3.5* Communication**	***8.3.5* Communication**		✔				Subclause heading number change.
***7.3.6* Tracking and escalation**	***8.3.6* Tracking and escalation**		✔				Subclause heading number change.
***7.3.7* Incident and problem record closure**	***8.3.7* Incident and problem record closure**		✔				Subclause heading number change.
***7.3.8* Problem reviews**	***8.3.8* Problem reviews**		✔				Subclause heading number change.

Table 7 – The Part 2 differences (continued)

BS 15000-2	ISO/IEC 20000-2	N/C	Form	Edit	Rec	Req	Commentary
Problem reviews are typically:	Problem reviews are typically:	✔					No change, left in for context of other changes.
a) reviews of individual incident levels and problem status against service levels;	a) reviews of individual incident levels and problem status against service levels;	✔					
b) management reviews to highlight those problems that require immediate action;	b) management reviews to highlight those problems that require immediate action;	✔					
c) management review to determine and analyse trends and to provide input for other processes, such **as customer education.**	c) management review to determine and analyse trends and to provide input for other processes, such **as user education and training.**			✔			Clarification from standardization of terms.
	8.3.9 Topics for reviews		✔				Subclause heading added as part of ISO formatting.
The reviews should include identification of:	The reviews should include identification of:	✔					
1) trends, e.g. recurring problems and incidents, known errors etc;	a) trends, e.g. recurring problems and incidents, known errors etc;		✔				Change form numbered bullets to alpha bullets, for ISO/IEC formatting.
2) recurring problems of a particular classification component or location;	b) recurring problems of a particular classification component or location;		✔				
3) deficiencies caused by resourcing, training or documentation;	c) deficiencies caused by resourcing, training or documentation;		✔				
4) **nonconformances**, e.g. against standards, policies and legislation;	d) **non-conformances**, e.g. against standards, policies and legislation;		✔				
5) known errors in planned releases;	e) known errors in planned releases;		✔				
6) staff resource commitment in resolving incidents and problems;	f) staff resource commitment in resolving incidents and problems;		✔				
7) recurrence of resolved incidents or problems.	g) recurrence of resolved incidents or problems.		✔				

Table 7 – The Part 2 differences (continued)

BS 15000-2	ISO/IEC 20000-2	N/C	Form	Edit	Rec	Req	Commentary
Improvements to the service or the problem management process should be recorded and fed **into a service improvement plan for action.**	Improvements to the service or the problem management process should be recorded and fed **into a plan for improving the service.**			✔			Clarification from standardization of terms.
The information should be added to the problem management knowledge base.	The information should be added to the problem management knowledge base.	✔					No change, left in for context of other changes.
All relevant documentation should be updated, e.g. user guides and system documentation.	All relevant documentation should be updated, e.g. user guides and system documentation.	✔					No change, left in for context of other changes.
7.3.9 ***Problem prevention***	***8.3.10*** ***Problem prevention***		✔				Subclause heading number change as part of ISO/IEC formatting.
8 Control processes	**9 Control processes**		✔				Clause heading number change, aligning Parts 1 and 2.
8.1 **Configuration management**	**9.1** **Configuration management**		✔				Clause heading number change, aligning Parts 1 and 2.
8.1.1 ***Configuration management planning and implementation***	***9.1.1*** ***Configuration management planning and implementation***		✔				Subclause heading number change as part of ISO/IEC formatting.
Configuration management should be planned and implemented with change and release management to ensure that the **organization** can manage its IT assets and configurations effectively.	Configuration management should be planned and implemented with change and release management to ensure that the **service provider** can manage its IT assets and configurations effectively.			✔			Clarification from standardization of terms.
Accurate configuration information should be available to support the planning and control of changes as new and updated services and systems are released and distributed.	Accurate configuration information should be available to support the planning and control of changes as new and updated services and systems are released and distributed.	✔					No change, left in for context of other changes.

Table 7 – The Part 2 differences (continued)

BS 15000-2	ISO/IEC 20000-2	N/C	Form	Edit	Rec	Req	Commentary
The result should be an efficient system that integrates the <u>organization's</u> configuration information processes and those of its customers and suppliers, where appropriate.	The result should be an efficient system that integrates the <u>service provider's</u> configuration information processes and those of its customers and suppliers, where appropriate.			✓			Clarification from standardization of terms.
<u>8.1.2</u> Configuration identification	*<u>9.1.2</u> Configuration identification*		✓				Subclause heading number change as part of ISO/IEC formatting.
	NOTE Other items that may be considered as configuration items include: a) other documentation; b) other assets; c) other facilities, e.g. site; d) business units; e) people.		✓				Moved from two sentences later in the text and converted to a NOTE as part of ISO formatting. This is not change to either the requirements or the recommendations.
Appropriate relationships and dependencies between configuration items should be identified to provide the necessary level of control.	Appropriate relationships and dependencies between configuration items should be identified to provide the necessary level of control.	✓					No change, left in for context of other changes.
Where traceability is required the process should ensure that configuration items can be traced through the full lifecycle, from requirements documents through to release records, e.g. using a traceability matrix.	Where traceability is required the process should ensure that configuration items can be traced through the full lifecycle, from requirements documents through to release records, e.g. using a traceability matrix.	✓					No change, left in for context of other changes.
Other items that may be considered as configuration items include: 1) other documentation; 2) other assets; 3) other facilities, e.g. site; 4) business units; 5) people.			✓				Moved to become a NOTE, two sentences earlier in the clause. Wording unchanged.

Table 7 – The Part 2 differences (continued)

BS 15000-2	ISO/IEC 20000-2	N/C	Form	Edit	Rec	Req	Commentary
<u>8.1.3</u> Configuration control	***<u>9.1.3</u> Configuration control***		✔				Subclause heading number change as part of ISO/IEC formatting.
<u>8.1.4</u> Configuration status accounting and reporting	***<u>9.1.4</u> Configuration status accounting and reporting***		✔				Subclause heading number change as part of ISO/IEC formatting.
Reports should cover:	Reports should cover:	✔					No change, left in for context of other changes.
a) latest configuration item versions;	a) latest configuration item versions;	✔					
b) location of the configuration item and for software the location of the master versions;	b) location of the configuration item and for software the location of the master versions;	✔					
c) interdependencies;	c) interdependencies;	✔					
d) version history.	d) version history;	✔					
<u>At any given time it should be possible to collate the status of configuration items that together constitute:</u>	**<u>e) status of configuration items that together constitute:</u>**			✔			Change to wording as part of ISO/IEC formatting. This is not a change to the original intent of these recommendations.
1) service configuration or system;	1) service configuration or system;			✔			
2) a change, baseline, build or release;	2) a change, baseline, build or release;			✔			
3) version or variant.	3) version or variant.			✔			
<u>8.1.5</u> Configuration verification and audit	***<u>9.1.5</u> Configuration verification and audit***		✔				Subclause heading number change as part of ISO/IEC formatting.
Configuration verification and audit processes, both physical and functional, should be scheduled and a check performed to ensure that adequate processes and resources are in place to:	Configuration verification and audit processes, both physical and functional, should be scheduled and a check performed to ensure that adequate processes and resources are in place to:	✔					No change, left in for context of other changes.

Table 7 – The Part 2 differences (continued)

BS 15000-2	ISO/IEC 20000-2	N/C	Form	Edit	Rec	Req	Commentary
a) protect the physical configurations and the intellectual capital of the organization;	a) protect the physical configurations and the intellectual capital of the organization;	✔					No change, left in for context of other changes.
b) ensure that the **organization** is in control of its configurations, master copies and licences;	b) ensure that the **service provider** is in control of its configurations, master copies and licences;			✔			Clarification from use of a standard term.
c) provide confidence that configuration information is accurate, controlled and visible;	c) provide confidence that configuration information is accurate, controlled and visible;	✔					No change, left in for context of other changes.
d) ensure that a change, a release, a system or an environment conforms to its contracted or specified requirements and that the configuration records are accurate.	d) ensure that a change, a release, a system or an environment conforms to its contracted or specified requirements and that the configuration records are accurate.	✔					
Configuration audits should be carried out regularly, before and after major change, after a disaster and at random intervals.	Configuration audits should be carried out regularly, before and after major change, after a disaster and at random intervals.	✔					
Deficiencies and non-conformities should be recorded, assessed and corrective action **instigated**, ...	Deficiencies and non-conformities should be recorded, assessed and corrective action **initiated**, ...			✔			This is not a change to the recommendations. The wording has been changed to clarify the meaning for an international audience.
..... acted upon and fed back to the relevant parties **and service improvement programme.**	..... acted upon and fed back to the relevant parties **and plan for improving the service.**			✔			Clarification from standardization of terms.
8.2 Change management	**9.2 Change management**		✔				Clause heading number change, aligning Parts 1 and 2.
Objective: To ensure all changes are assessed, approved, implemented and reviewed in a controlled manner.	*Objective: To ensure all changes are assessed, approved, implemented and reviewed in a controlled manner.*	✔					

Table 7 – The Part 2 differences (continued)

BS 15000-2	ISO/IEC 20000-2	N/C	Form	Edit	Rec	Req	Commentary
8.2.1 Planning and implementation	*9.2.1 Planning and implementation*		✔				Subclause heading number change, aligning Parts 1 and 2.
The change management processes and procedures should ensure that:	The change management processes and procedures should ensure that:	✔					No change, left in for context of other changes.
a) changes have a clearly defined and documented scope;	a) changes have a clearly defined and documented scope;	✔					
b) only changes that provide business benefit are approved, e.g. commercial, legal, regulatory, statutory;	b) only changes that provide business benefit are approved, e.g. commercial, legal, regulatory, statutory;	✔					
c) changes are scheduled based on priority and risk;	c) changes are scheduled based on priority and risk;	✔					
d) changes to configurations can be verified during change implementation;	d) changes to configurations can be verified during change implementation;	✔					
e) the time to implement changes is monitored and improved where required.	e) the time to implement changes is monitored and improved where required	✔					
The change management process should be able to demonstrate how a change is:	f) it can be demonstrated how a change is:		✔				Change to formatting and structure that reflects the original intent of recommendations and aligns with ISO/IEC formatting practices.
1) raised, recorded and classified (with references to documents that gave rise to the change);	1) raised, recorded and classified (with references to documents that gave rise to the change);		✔				
2) assessed for the impact, urgency, cost, benefits and risk of the changes on the service, customer and release plans;	2) assessed for the impact, urgency, cost, benefits and risk of the changes on the service,		✔				
customer and release plans;	3) reversed or remedied if unsuccessful;		✔				

Table 7 – The Part 2 differences (continued)

BS 15000-2	ISO/IEC 20000-2	N/C	Form	Edit	Rec	Req	Commentary
4) documented, e.g. the change request is linked to affected configuration items and the updated version of the implementation and release plans;	4) documented, e.g. the change request is linked to affected configuration items and the updated version of the implementation and release plans;		✔				
5) approved or rejected by a change authority, depending on the type, size and risk of change;	5) approved or rejected by a change authority, depending on the type, size and risk of change;		✔				
6) be implemented by the nominated owner within the groups responsible for the components being changed;	6) be implemented by the nominated owner within the groups responsible for the components being changed;		✔				
7) tested, verified and signed off;	7) tested, verified and signed off;		✔				
8) closed and reviewed;	8) closed and reviewed;		✔				
9) scheduled, monitored and reported on;	9) scheduled, monitored and reported on;		✔				
10) linked to incident, problem, other change and configuration item records where appropriate.	10) linked to incident, problem, other change and configuration item records where appropriate.		✔				
The status of changes and scheduled implementation dates should be used as the basis for change and release scheduling.	The status of changes and scheduled implementation dates should be used as the basis for change and release scheduling.	✔					
Scheduling information should be made available to the people affected by the change.	Scheduling information should be made available to the people affected by the change.	✔					No change, left in for context of other changes.
Where an outage can be caused during normal service hours the people affected should agree to the change before implementation.	Where an outage can be caused during normal service hours the people affected should agree to the change before implementation.	✔					

Table 7 – The Part 2 differences (continued)

BS 15000-2	ISO/IEC 20000-2	N/C	Form	Edit	Rec	Req	Commentary
<u>8.2.2</u> Closing and reviewing the change request	***<u>9.2.2</u> Closing and reviewing the change request***		✔				Subclause heading number change.
Any weaknesses or deficiencies identified in a review of the **<u>change control process</u>** …..	Any weaknesses or deficiencies identified in a review of the **<u>change management process</u>** …..			✔			Clarification from standardization of terms. This is not a change to the original intent of this recommendation in BS 15000-2, but the original wording may have been mis-understood.
…..should be fed **<u>in to service improvement plans.</u>**	….. should be fed **<u>into plans for improving the service.</u>**			✔			Clarification from standardization of terms.
<u>8.2.3</u> Emergency changes	***<u>9.2.3</u> Emergency changes***		✔				Subclause heading number change as part of ISO/IEC formatting.
<u>8.2.4</u> Change management reporting, analysis and actions	***<u>9.2.4</u> Change management reporting, analysis and actions***		✔				Subclause heading number change as part of ISO/IEC formatting.
9 Release management processes	**10 Release process**		✔				No change, left in for context of other changes.
	10.1 Release management process		✔				New heading inserted for alignment of Parts 1 and 2.
	10.1.1 General		✔				New subclause heading inserted for ISO formatting.
Good planning and management are essential to package and successfully distribute a release, and to manage the associated impact and risks to the business and IT.	Good planning and management are essential to package and successfully distribute a release, and to manage the associated impact and risks to the business and IT.	✔					No change, left in for context of other changes.

Table 7 – The Part 2 differences (continued)

BS 15000-2	ISO/IEC 20000-2	N/C	Form	Edit	Rec	Req	Commentary
The release of affected information systems, infrastructure, services and documentation should be planned with the business.	The release of affected information systems, infrastructure, services and documentation should be planned with the business.	✔					No change, left in for context of other changes.
Release management should be integrated with the configuration and change management processes.					✔		Moved to Part 1 and changed to a NOTE. This is a change to recommendations in Part 2, but the NOTE in Part 1 has the same meaning as this sentence in Part 2.
All associated updates to documentation should be included in the release, e.g. business processes, support documents and service level agreements.	All associated updates to documentation should be included in the release, e.g. business processes, support documents and service level agreements.	✔					No change, left in for context of other changes.
9.2 Release policy	***10.1.2 Release policy***		✔				Change to subclause heading level aligning Parts 1 and 2.
9.3 Release and roll-out planning	***10.1.3 Release and roll-out planning***		✔				Change to subclause heading level aligning Parts 1 and 2.
The service provider should work with the business to ensure that the configuration items **that are to be released are compatible.**	The service provider should work with the business to ensure that the configuration items **that are to be released are compatible with each other and with configuration items in the target environment.**			✔			Clarification from re-wording. This is not a change to the intent of this clause, which may have been mis-interpreted.
9.4 Developing or acquiring software	***10.1.4 Developing or acquiring software***		✔				Change to subclause heading level aligning Parts 1 and 2.
9.5 Design, build and configure release	***10.1.5 Design, build and configure release***		✔				Change to subclause heading level aligning Parts 1 and 2.
Release and distribution should be designed and implemented to:	Release and distribution should be designed and implemented to:	✔					No change, left in for context of other changes.

Table 7 – The Part 2 differences (continued)

BS 15000-2	ISO/IEC 20000-2	N/C	Form	Edit	Rec	Req	Commentary
a) conform with the **organization's** systems architecture, service management and infrastructure standards;	a) conform with the **service provider's** systems architecture, service management and infrastructure standards;			✔			Clarification from standardization of terms.
9.6 Release verification and acceptance	***10.1.6 Release verification and acceptance***		✔				Change to subclause heading level aligning Parts 1 and 2.
The end result should be a sign-off on completeness of the whole release package against requirements.	The end result should be a sign-off on completeness of the whole release package against requirements.	✔					No change, left in for context of other changes.
	10.1.7 Documentation		✔				Change to subclause heading level aligning Parts 1 and 2.
Appropriate documentation should be available on completion and stored under configuration management against the released configuration item.	Appropriate documentation should be available on completion and stored under configuration management against the released configuration item.	✔					No change, left in for context of other changes.
This documentation should include:	This documentation should include:		✔				Minor formatting change, alpha bullets instead of numeric.
9.7 Roll-out, distribution and installation	***10.1.8 Roll-out, distribution and installation***		✔				Change to subclause heading level aligning Parts 1 and 2.
9.8 Post release and roll-out	***10.1.9 Post release and roll-out***		✔				Change to subclause heading level aligning Parts 1 and 2.
Recommendations should be fed into **the service improvement plan.**	Recommendations should be fed into **a plan for improving the service.**			✔			Clarification form standardization of terms.

Table 7 – The Part 2 differences (continued)

BS 15000-2	ISO/IEC 20000-2	N/C	Form	Edit	Rec	Req	Commentary
Bibliography	**Bibliography**		✔				
	[1] ISO/IEC 20000, Information technology - service management — Part 1: Specification		✔				With this one exception the Bibliography is identical to that in Part 1 and completely replaced that in BS 15000.
	National Annex		✔				
	See Part 1		✔				National Annex is identical to that in Part 1.
	National bibliography		✔				
	See Part 1		✔				National bibliography is identical to that in Part 1.

Appendix A

Bibliography and further information

Standards

BS 0-3, *A standard for standards — Part 3: Specification for structure, drafting and presentation*

ISO/IEC Directives Part 2, *Rules for the structure and drafting of International Standards*

ISO 9000, *Quality management systems — Fundamentals and vocabulary*

ISO 9001, *Quality management systems — Requirements*

ISO/IEC 17799, *Information technology — Security techniques — Code of practice for information security management*

ISO/IEC 20000-1, *Information technology — Service management — Part 1: Specification*

ISO/IEC 20000-2, *Information technology — Service management — Part 2: Code of practice*

ISO/IEC 27001, *Information technology — Security techniques — Information security management systems — Requirements*

BSI books

BIP 0005, *A manager's guide to service management*

BIP 0015, *IT service management — Self-assessment workbook*

BIP 0008, *Code of practice for legal admissibility and evidential weight of information stored electronically*

PAS 56, *Guide to business continuity management*

Security information

BIP 0070, *Information security compilation on CD-ROM*

BIP 0071, *Guidelines on requirements and preparation for certification based on ISO/IEC 27001*

BIP 0072, *Are you ready for an ISMS audit based on ISO/IEC 27001?*

BIP 0073, *Guide to the implementation and auditing of ISMS controls based on ISO/IEC 27001*

BIP 0074, *Measuring the effectiveness of your ISMS implementations based on ISO/IEC 27001*

Other resources

British Computer Society: www.bcs.org.uk

British Computer Society Configuration Management Specialist Group: www.bcs-cmsg.org.uk

The IT Service Management Forum (itSMF): www.itsmf.com

EXIN: www.exin.nl

Information Systems Examinations Board (ISEB): www.bcs.org.uk/iseb

The Office of Government Commerce: www.ogc.gov.uk

IT Infrastructure Library (ITIL): www.itil.co.uk

BOOKS IN THE 'ACHIEVING ISO/IEC 20000' SERIES

There are ten books in the 'Achieving ISO/IEC 20000' series. Each book in the series includes an abstract of ISO/IEC 20000 that is most relevant to the topic of the book, as well as useful contacts and sources of supporting information. These books can be purchased through the BSI website at www.bsi-global.com.

BIP 0030, *Management decisions and documentation*

This book covers: the background to ISO/IEC 20000; a comparison to other standards and best practice material; compliance and certification audits; the scope of service management; building the business case for achieving ISO/IEC 20000; preparation for an audit and using ISO/IEC 20000 to select your supplier. Important terms that are used in management system standards, where the exact meaning of terms is important to the correct interpretation of the standard, are also explained, including the differences between the terms '**shall**', '**should**' and notes. This book also covers the requirements and recommendations for documents and records, which is a management responsibility requirement in clause 3.2 of ISO/IEC 20000-1.

BIP 0031, *Why people matter*

This book covers the roles and responsibilities of management and process owners, and explains the importance of management commitment to best practice service management, mapping onto the requirements and recommendations of clause 3.1 of ISO/IEC 20000, *Management responsibility*. The book also covers the importance of motivation, training and career development as well as tips and techniques, mapping onto the requirements of clause 3.3 of ISO/IEC 20000-1, *Competence, awareness and training*.

BIP 0032, *Making metrics work*

This book gives a practical view of why metrics and service reports are so important to the delivery of an effective service and to service improvements. It describes the types, the design, target audiences and documentation of metrics used in the service reporting process, covered by the requirements of clauses 4 and 6.2 of ISO/IEC 20000-1, *Plan-Do-Check-Act (PDCA) cycle* and *Service reporting*. Useful tips, techniques and example metrics are included.

BIP 0033, *Managing end-to-end service*

This book describes the supply chains that are commonly managed by service level management, business relationship management and supplier management, which are the requirements in clauses 6.1 and 7 of ISO/IEC 20000-1. It describes the interfaces between suppliers, the service provider and one or many customers. This book also includes useful tips for aspects of end-to-end service, such as the role of service level agreements (SLAs), service reviews, customer satisfaction and complaint handling.

BIP 0034, *Finance for service managers*

This book covers *Budgeting and accounting for IT services* based on clause 6.4 of ISO/IEC 20000. It introduces financial terms that may be unfamiliar to service management specialists, which will help with understanding the requirements and recommendations. It also covers the relationship between budgeting, accounting and charging, and outlines the importance of service management processes in regulatory compliance.

BIP 0035, *Enabling change*

This book covers the configuration, change management and release management processes which are contained in clauses 9 and 10 of ISO/IEC 20000. It compares the three processes and describes how they interface with each other, and gives advice on the requirements and recommendations of ISO/IEC 20000, example metrics and audit evidence. This book also includes practical advice on meeting the ISO/IEC 20000 requirements on the roles and responsibilities of those involved.

BIP 0036, *Keeping the service going*

This book covers the service continuity and availability management, incident management and problem management processes, which are contained in clauses 6.3 and 8 of ISO/IEC 20000. It explains the role of these processes in keeping the customer's service going, ranging from continuity planning through to the fast-fixing of incidents. It compares the processes and describes how they interface with each other. It includes example metrics and audit evidence, with practical tips and techniques that will help a service provider achieve the requirements.

BIP 0037, *Capacity management*

This book covers the requirements for the capacity management process in clause 6.5 of ISO/IEC 20000. It describes the capacity management process and its role as a link between business plans, workloads, capacity and performance). It also covers the planning required to ensure a service provider is able to deliver a service that allows the customer's business to operate effectively. The book describes capacity management for all types of resources within the scope of service management.

BIP 0038, *Integrated service management*

The opening paragraph of ISO/IEC 20000-1 states that '*This standard promotes the adoption of an integrated process approach to effectively deliver managed services to meet the business and customer requirements*'. This book reflects the importance placed by ISO/IEC 20000 on understanding the interfaces between processes, and how the interfaces are managed so that service management processes are fully integrated. It also reflects the top-down management system approach that is fundamental to ISO/IEC 20000. This book describes how understanding and meeting the requirements of ISO/IEC 20000 gives better control, greater efficiency and opportunities for improvements.

BIP 0039, *The differences between BS 15000 and ISO/IEC 20000*

This book will be of particular interest to those who have used BS 15000 for service improvements, audits or training and need to update their material to reflect the ISO/IEC 20000 standard. ISO/IEC 20000 was based on BS 15000, and this book provides a detailed comparison of ISO/IEC 20000 and BS 15000, for both Parts 1 and 2. It shows the differences in structure, clause numbering and references. The core of this book is a series of tables detailing the changes to the requirements and recommendations clause-by-clause, as well as any re-wording that has been provided to give clarification for an international audience. It includes an explanation of why the changes were made and the implications of each of the changes.
This book is based on the material produced by the Project Editor during the drafting of both Parts 1 and 2 of ISO/IEC 20000.